Total
Productive
Maintenance
An American Approach

Terry Wireman

Total Productive Maintenance

An American Approach

Industrial Press Inc.

Library of Congress Cataloging-in-Publication Data

Wireman, Terry.
   Total productive maintenance : an American approach / Terry
Wireman.
     192 p. 15.6 × 23.5 cm.
     Includes index.
     ISBN 0-8311-3036-9 :
     1. Plant maintenance—United States—Management.
  2. Industrial equipment—United States—Maintenance and repair
—Management.
  I. Title.
TS192.W575   1991
658.2'7—dc20   91-16772
                                                             CIP

Industrial Press Inc., 200 Madison Avenue, New York, NY 10016

First Edition

Composition by Pub-Set, Inc., Union, NJ.

Printed and bound by Quinn Woodbine, Woodbine, NJ.

10 9 8 7 6 5 4 3 2

Dedication

This book is dedicated to my wife Pam and my sons Justin and Chad. It was only through their patient understanding that this book was ever finished.

Preface

This book presents a modification of many of the standard maintenance techniques and policies that are currently used in the United States. It is only fitting that an American version of Total Productive Maintenance should be developed. After all, it was modifying American techniques that led to the Japanese version of T.P.M. Turnabout should be fair play.

While many companies already practice some of the disciplines described in this book, very few take the time and see the whole picture. It is only by taking the objective approach to maintenance improvement that the real improvements will be identified and made. Without trying to understand how maintenance can be controlled and managed effectively, many companies sacrifice a tremendous amount of money to old ideas.

The old ideas include the belief that maintenance is a necessary evil, that it is just overhead, or that it is a continual cash drain. It must be remembered that the same things were said about quality programs a decade ago. Yet, how many companies would say that about their quality, considering the present attitude toward quality in World Class Manufacturing? The same transformation will take place with maintenance in the decade of the 90's.

Total Productive Maintenance—An American Approach is designed to help organizations optimize the effective use of corporate assets in order to maintain a competitive advantage in their respective marketplaces. In the quote "Automate, Innovate or Evaporate," maybe some ideas on how to innovate can be gleaned from this material to help your company survive.

In rounding out the preface, I would like to thank Neil Drummond and Steve Pettit for their help in providing occasional sounding boards for some of the concepts contained in this book. Their expertise is recognized and respected by any company that has ever had the opportunity to work with them.

Contents

Introduction:
Wasteful Practices, and
Why Maintenance is Important

When the topic of maintenance is brought up in management meetings, it conjures up differing visions for individuals. The operations manager may envision an insurance policy; a plant manager may visualize a firefighter; a division manager may mentally picture unnecessary overhead. Unfortunately, few managers properly picture maintenance departments as major contributors to corporate profitability.

Studies have revealed that maintenance costs are between 15 and 40% of the total cost of goods for typical manufacturing. A savings in maintenance expenditures adds directly to the corporate bottom line. In addition, other studies have shown that 30% of all maintenance expenditures are wasted. Also consider: "What percentage of a company's capital assets are invested in backup or redundant systems?" Many corporations invest in such systems to ensure their ability to deliver the product as scheduled. They do this since, in some companies, the equipment availability is barely over 50%. If proper maintenance techniques were practiced (perhaps raising availability to 90–95%), the capital investment could be reduced, allowing the company to invest the assets in a more profitable manner. A final point for consideration is the actual cost for equipment unavailability or downtime. Surveys show that the cost for a breakdown or outage is far more than the maintenance labor and materials to effect the repair. One survey showed that the actual cost for a breakdown averaged 4 times the maintenance cost; in some companies, it was 15 times the maintenance cost. When examining the actual impact maintenance has on corporate profitability, is the true cost really considered? If all of these points for consideration are combined, it can truly be said that maintenance may be the last major cost savings frontier for management.

How significant is this contribution that maintenance can make to corporate profitability? It is important to note that the definition of World Class Competitiveness indicates that the qualifications are to provide the highest quality product or service, and to be able to provide it on time and at the lowest possible cost. By this definition, if a competitor pays less of his cost of sales for maintenance, while producing an equal quality product, he can offer it for a lower price and still maintain his profit margin. This makes his company World Class, and makes the company with poor maintenance practices second rate.

In addition to the points already illustrated, also consider how maintenance impacts cost in the following areas.

1) **Quality**—What percent of all quality problems is eventually attributed to poor maintenance policies or practices? How many quality problems must be corrected by maintenance?

2) **Warranty Costs**—What percent of equipment repairs is performed on equipment that is under warranty? What part of these expenses is recoverable from the vendor?

3) **Energy Costs**—It is a fact that poorly maintained equipment requires more energy to operate.

4) **Accelerated Depreciation of Capital Assets**—What is the amount of equipment/facility replacement that is performed prematurely, due to poor maintenance policies and practices?

In view of the above issues, it is obvious that maintenance has a greater impact on corporate profitability than most managers are willing to consider, much less admit. As the competitive environment in the world continues to increase the pace, how much longer can companies fail to consider the maintenance factor?

Now we introduce the term from Japan—Total Productive Maintenance. It is a fact that companies are considering this method to take advantage of the cost avoidance (profit) that optimum maintenance policies and practices provide. If other world competitors are using this tool, can anyone afford not to be doing the same? But just what is T.P.M.? How would you implement it in your company? How would you convince people of the benefits it can achieve? It is the purpose of this text to explore those questions and more. But beyond that, this text will provide a practical, useful implementation plan that can be adjusted and modified to allow any company to achieve the maximum competitive advantage from its maintenance organization.

Preparing for Total Productive Maintenance

In Part I, we will examine the activities necessary for a company to prepare for T.P.M. This will include a clear understanding of what T.P.M. really is, and what it is not. We will examine the current state of maintenance management in the United States, while illustrating why improvement is vitally important to corporations. We will also highlight preimplementation activities that must be conducted before any T.P.M. program can be successful.

1 The History and Impact of Total Productive Maintenance

In order to properly understand the history and impact of Total Productive Maintenance, it is necessary to establish its definition. T.P.M. is maintenance that involves all employees in the organization; this means everyone from top management to the line employee. It encompasses all departments including:

maintenance

operations

facilities

design engineering

project engineering

construction engineering

inventory and stores

3

purchasing

accounting finances

plant/site management.

The goal of T.P.M. can be divided into the following five basic segments:

1. ensure equipment capacity,

2. implement a program of maintenance for the entire life of the equipment,

3. require support from all departments involved in the use of the equipment or facility,

4. solicit input from all employees from all levels of the company, and

5. use consolidated teams for continuous improvement.

1) Ensuring equipment capacity indicates that the equipment performs to specifications. It operates at its design speed, produces at the design rate, and results in a quality product at these speeds and rates. The problem with the equipment is that many companies do not even know the design speed or rate of production. This allows management to set arbitrary production quotas. A second problem is that, over time, small problems cause operators to change the rate at which they run equipment. As these problems continue to build, the equipment output may only be 50% of what it was designed for. This will lead to the investment of additional capital in equipment, trying to meet the required production output.

2) Implementing a program of maintenance for the life of the equipment is analogous to the popular preventive and predictive maintenance programs companies presently use to maintain

maintenance programs companies presently use to maintain their equipment. This step has a basic difference: the program changes just as the equipment changes. Each piece of equipment requires different amounts of maintenance as it ages. A good preventive/predictive maintenance program takes these changing requirements into consideration. By monitoring failure records, trouble calls, and basic equipment conditions, the program is modified to meet the changing needs of the equipment.

A second difference is that this program involves all employees, from the operator to upper management. The operator may be required to perform basic cleaning and lubricating of the equipment, which is really the front line defense against problems. The upper managers may be required to assure that maintenance gets enough time to properly finish any service or repairs required to maintain the equipment in the condition necessary to keep it running at designed ratings.

3) Requiring the support of all departments involved in the use of the equipment or facility will ensure full cooperation and understanding of affected departments. For example, including maintenance in equipment design/purchase decisions assures that equipment standardization is considered. The issues surrounding this topic alone can contribute significant financial savings to the company. Standardization reduces inventory levels, training requirements, and start-up times. Another area is the support given to maintenance from stores and purchasing. Good support can help reduce downtime, but more important is the optimization of inventory levels to prevent carrying too much inventory.

4) Soliciting input from employees at all levels of the company provides employees with the ability to contribute to the process. In most companies, this step takes the form of a suggestion program. However, it needs to go beyond that; it should include a management with "no doors." This indicates that man-

agers, from the front line to the top, must be open and available to listen to and give consideration to employee suggestions. A step further is the response that should be given to each discussion. It is no longer sufficient to say "That won't work" or "We are not considering that now." In order to keep communication flowing freely, reasons must be given. It is just a matter of developing and utilizing good communication and management skills. Without these skills, employee input will be destroyed, and the ability to capitalize on the greatest savings generator in the company will be lost.

5) The development of the consolidated teams for continuous improvement begins with step 4. The more open management is to the ideas of the workforce, the easier it is for the teams to function. These teams can be formed by areas, departments, lines, process, or equipment. They will involve the operators, maintenance, and management personnel. They will, depending on the needs, involve other personnel on an as-needed basis, such as engineering, purchasing, or stores. These teams will provide answers to problems that some companies have tried years to solve independently. This team effort is one of the true indicators of a successful T.P.M. program.

The questions now raised are: "Is it all worth it?" "What are the benefits that have been achieved?" These questions are answered positively and quickly as evidenced by the following results.

Productivity:
    100–200% increases
    50–100% increase in rates of operation
    500% decrease in breakdowns.

Quality:
    100% decrease in defects
    50% decrease in client claims.

Costs:

    50% decrease in labor costs

    30% decrease in maintenance costs

    30% decrease in energy costs.

Inventory:

    50% reduction of inventory levels

    100% increase in inventory turns.

Safety:

    Elimination of environmental and safety violations.

Morale:

    200% increase in suggestions

    increased participation of employees in small group meetings.

With all of these benefits, it is important that all companies recognize the importance and value that maintenance, specifically productive maintenance, can bring to the company. Any company trying to achieve "World Class" stature through other programs such as CIM, JIT, TQC, or TEI will soon find that they will not work without that total reliability of the company's assets, which is the primary responsibility of the maintenance organization. A good analogy is the cutlery for a meal. It is difficult to eat a full meal without all three utensils—knife, fork, and spoon. Striving for "World Class" status is the same. Just-in-Time, Total Quality Control, and Total Productive Maintenance are all essential if the goal is to be reached. Without full utilization of these three programs, the goal of being globally competitive will never be reached.

As the decade of the '90's begins, the focus on T.P.M. is intensifying. Why the interest? It only takes a quick review of the current state of maintenance in the U.S. to see that changes are required if companies want to be competitive.

Maintenance Costs

It is estimated that U.S. companies will have spent over 600 billion dollars on maintenance and related expenditures in 1990. This is a tremendous dollar amount, but is even more alarming when approximately 1/3 of it is unnecessary or wasted. This is an advantage in costs that companies can ill afford to give to their foreign competitors.

Where are the wastes? They are in the ineffective use and control of maintenance resources, labor, and materials. For example, what percent of the time is a maintenance technician involved in actual hands-on activities? Is it 2 or 3 hours out of 8? In companies where reactive or emergency types of maintenance make up 50% or more of the maintenance workload, technicians average 2–3 hours of hands-on activities. The rest of the time, they are engaged in nonproductive activities such as looking for parts, drawings, instructions, or authorization.

What about inventory wastes? The cost of having too many spares is paid, not only in capital investments, but also in carrying costs, storage costs, and labor costs. This does not count the spoilage costs, the pilferage costs, or the costs of damaged materials being stored and moved frequently.

A survey of maintenance and maintenance-related personnel recently showed that the following areas needed work:

maintenance scheduling,

hiring and training maintenance technicians,

too much emergency or breakdown maintenance,

lack of controls over maintenance spares, and

lack of upper management support and understanding.

Each of these problems is difficult to solve, but when combined they provide any manager with a formidable task. However, organizations that are having these problems will have an almost impossible task to implement a T.P.M. program. The right step would be to solve some of these basic problems before tackling the task of implementing T.P.M. Later in this book, we will explain how to solve these problems.

Maintenance Budgets

Maintenance budgeting is another problem for organizations. There are many methods used to budget and monitor maintenance. Some work well, others are burdens to the maintenance departments. The extremes are where maintenance is responsible for all maintenance-incurred expenses, whether or not they are requested or approved by maintenance or operations. This means they may or may not be in control of the monies being spent. The opposite end of the spectrum is the organization with zero maintenance budget. All charges, including an overhead multiplier, are billed directly to the department which requests the work or "owns" the equipment. In this case, the operations or facility organization will want to keep the maintenance costs as low as possible, and will defer repairs/improvements/refurbishments that should have been performed. In either case, the primary consideration, which is the condition of the company's capital assets (the equipment or facility) has been pushed into the background. Total Productive Maintenance focuses on the equipment/facility, and this pays tremendous benefits to the company.

The cost/payback for maintenance versus the company cost is shown Fig. 1-1. Maintenance costs are between 15 and 40% of the total cost of production in typical manufacturing. The average is approximately 28%. This is high, when considering that

MANUFACTURING COSTS

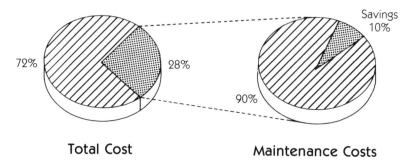

Total Cost   **Maintenance Costs**

Figure 1-1. Reducing maintenance costs by 10% can produce an increase in pre-tax profit of almost 36%.

pre-tax profit for typical manufacturing is 5%. If maintenance costs are reduced 10%, this cost avoidance goes directly to the pre-tax profit. This results in approximately a 36% increase in the pre-tax profits. This is a good reason for improving maintenance, but is only part of the savings. In some companies, a savings of as much as 50% of the maintenance budget has been reported. This savings increases the pre-tax profit considerably, and will allow the company to be even more competitive in its respective market. The comparative savings is highlighted in Fig. 1-2.

While this savings is enough of an enticement for some companies, the true cost savings is yet to come. Consider which is more: the maintenance cost of a repair or the lost production/ increased production costs? One survey showed that this cost ranges from 2:1 to as high as 15:1, as shown in Fig. 1-3. So if the maintenance cost for a repair was $10,000, was the true cost to the company $30,000 or $160,000? It is critical that companies examine the true cost of maintenance or nonmaintenance if they are ever to be successful in improving maintenance and implementing T.P.M.

PRE-TAX PROFIT

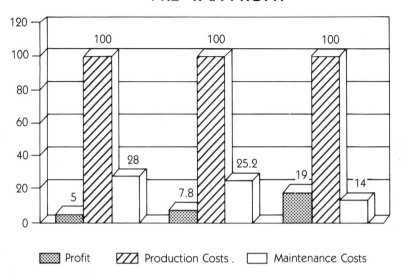

Figure 1-2. The relationship between reduced maintenance costs and increased pre-tax profit.

LOST PRODUCTION/MAINTENANCE COST COMPARISON

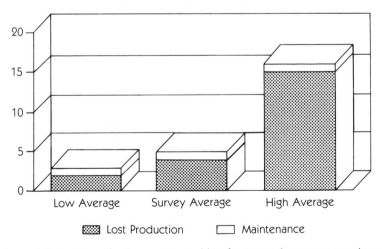

Figure 1-3. Lost production costs resulting from a maintenance repair can vary greatly in different companies.

An additional problem is the control of the maintenance budgeting process. In many organizations, maintenance managers do not even control their own budgets. In over 50% of sites, the plant manager or the plant engineer control the maintenance budget. This prevents the maintenance managers from controlling the department they are responsible for and held accountable to manage. This is highlighted in Fig. 1-4. Unless a manager is allowed to control his department's budget, he cannot be responsible for its effectiveness. Total Productive Maintenance places responsibility and control for the job functions with the correct managers.

Maintenance Control Systems

This problem area contributes to all aspects of maintenance and is a key ingredient to a successful T.P.M. program. This is the system used by maintenance to gather information and provide an engineering database to make accurate and cost-effective maintenance decisions. The common name for this system is the work order. Most companies will claim to have a work order

BUDGET RESPONSIBILITY

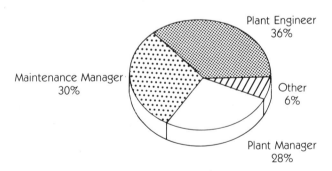

Figure 1-4. In only 30% of all sites does the maintenance manager control the maintenance budgeting process.

system, however, only a minority of the companies is satisfied with their information. This point is highlighted in Fig. 1-5. This figure shows that the basic information-gathering function in most maintenance organizations is not functioning properly. If the information is not being gathered properly, one must ask: *"how accurate are the decisions that are being made, based on this information?"*

Companies that accurately gather information on work order systems still fail to use it correctly. For example, of those companies that gather work order information, many fail to perform monitoring and information analysis (see Fig. 1-6). So even when some companies do gather the information, they fail to use it to find and implement cost-effective asset management decisions.

Even beyond the asset management issue, consider maintenance staff sizes. The size of a maintenance workforce is determined by the amount of work that it has to perform. This is commonly called the craft backlog. It is the accumulated total of all estimated labor requirements on work orders waiting to be performed. Since many companies do not plan and estimate

WORK ORDER SYSTEM ANALYSIS

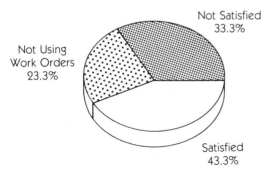

Figure 1-5. Approximately 56% of all companies are not using work orders, or are not satisfied with the information they provide.

WORK ORDER INFORMATION ANALYSIS

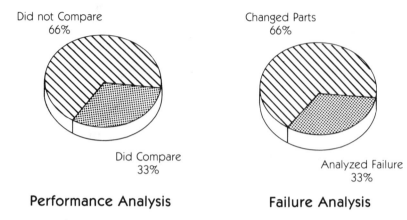

Did not Compare
66%

Changed Parts
66%

Did Compare
33%

Analyzed Failure
33%

Performance Analysis   **Failure Analysis**

Figure 1-6. Even though almost 77% of all companies use work orders, many of them fail to analyze the information they provide.

work orders, they do not know the size of their maintenance backlog(s). This is highlighted in Fig. 1-7. Since this is the case, how do they make justifiable decisions on maintenance staffing levels? It would be unimaginable for a production department to be managed in the same manner. You would either have operators standing around or equipment sitting idle without an operator. This area must be controlled if optimum use of resources is to be realized.

Preventive Maintenance

This is another major area that must be investigated during a T.P.M. development plan. How do U.S. companies perform preventive maintenance? Results indicate that almost 80% of the companies are not satisfied that their programs work or are cost effective. This is highlighted in Fig. 1-8. The main reasons for these failures, and the solutions, will be discussed in Chapter 6.

CRAFT BACKLOGS
MAN HOURS/CRAFT

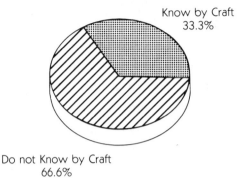

Know by Craft
33.3%

Do not Know by Craft
66.6%

Figure 1-7. Many companies fail to plan and estimate existing work orders.

However, the biggest reason for the failure of preventive/ predictive maintenance programs is the lack of understanding and support for the program by upper management. Total Productive Maintenance programs ensure this support. Preventive maintenance under a T.P.M. program will be successful, if it is properly designed and implemented.

P.M. PROGRAMS

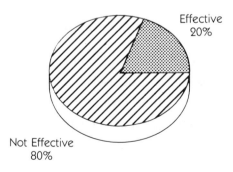

Effective
20%

Not Effective
80%

Figure 1-8. About 80% of all companies are not satisfied with their existing preventive maintenance programs.

Maintenance Inventory and Purchasing

This area is a major contributor to the lack of maintenance productivity in the United States. The time wasted while trying to find parts for maintenance technicians makes up one of the largest portions of the lost productive time previously discussed. It is estimated that over 50% of all lost maintenance productivity is related to inventory and purchasing practices.

This problem is even further compounded by the fact that in almost ½ of all companies, maintenance has no control over its stores and spares. This means that order policies and storage policies are made by individuals who may not understand how maintenance stores are different from operations or production stores. This creates stockouts or overages, which are both unnecessary expenses that weaken an organization's competitive position. While stockouts are considered a nuisance to maintenance, what is the true cost of a stockout? What is the cost of downtime or lost production that is created by a stockout? This can be considerable in some instances. However, it is rarely a factor considered in stocking decisions. This is a major flaw in U.S. maintenance stores.

Is the Japanese T.P.M. the Right Answer?

This is a question that many companies are asking themselves. The Japanese experts themselves admit that there may be problems implementing their style T.P.M. in American companies, due to culture, management, and labor difference. The main thrust of T.P.M. is the proper care and maintenance of the company's equipment and capital assets.

In review, the five main goals of T.P.M. are:

1. ensure equipment capacity,

2. implement a program of maintenance for the entire life of the equipment/facility/asset,

3. require support from all departments involved in the use of the equipment or facility,

4. solicit input from all employees from all levels of the company, and

5. use consolidated teams for continuous improvement.

It must now be stressed that there is no single right method for implementation of a T.P.M. program. The steps necessary to develop a T.P.M. program must be determined for each company

TEN-STEP T.P.M. PLAN

* Developing the Long-Range Plan

* Selling the Program

* Ensuring Equipment Reliability

* Maintenance Inventory Controls

* Improving Maintenance Efficiency

* Maintenance Automation

* Training for C.A.T.'s

* Optimizing Maintenance Resources

* Team-Based Maintenance

* Measuring the Results

Figure 1-9. The ten-step plan for implementing a total preventive maintenance program.

individually. The steps must be adjusted to fit individual requirements since the types of industry/service/facility, production methods, service activities, equipment conditions, special needs, problems, techniques, and levels of sophistication of maintenance vary dramatically from organization to organization. The focus of any T.P.M. program is the achievement of the 5 basic goals of T.P.M., and any others that may be identified by the particular organization.

The Ten-Step Plan

The second part of this book describes a ten-step implementation plan better suited for American-style companies. Again, it is an outline for the implementation; the specific steps may have to be altered or reprioritized for a successful implementation. The ten-step plan is shown in Fig. 1-9.

Before the ten-step plan can be understood, a further understanding of T.P.M. and some of the activities necessary before the implementation plan can be developed must be explained. These steps will be taken in the next two chapters.

2 Determining the Goals of Total Productive Maintenance

The true goals of Total Productive Maintenance revolve around the corporate assets: the equipment, the buildings and grounds, and the entire facility. Of the five basic segments, described fully in Chapter 1 and repeated here, ensuring equipment capacity is the most important.

1. Ensure equipment capacity,

2. Implement a program of maintenance for the entire life of the equipment,

3. Require support from all departments involved in the use of the equipment or facility,

4. Solicit input from all employees from all levels of the company, and

5. Use consolidated teams for continuous improvement.

19

Ensuring equipment capacity means that the equipment operates at its design specifications for the entire life of the equipment. The goal is to increase equipment effectiveness or capabilities by minimizing input and maximizing output. This goal applies to facilities as well as production plants.

In facilities, the input includes the labor resources, energy resources, and the supplies and miscellaneous items that are utilized at the site. For example, unless the physical plant's and building's heating, lighting, etc., are properly maintained, there is waste. When energy usage is higher than required, a corporate asset—money—is wasted. The building should be maintained at a level where the maximum benefit is attained from each resource used by the facility.

In production environments, input includes the resources required to produce the product. The output is

increased capacity and productivity

highest quality

lower production costs

on-time delivery

improved safety and hygiene

higher employee morale

a more favorable workplace environment.

These benefits are desirable to any company, but there are obstacles in the path to achieving them. The first major obstacle is the attitude of all personnel from top management to the line employee. It is the acceptance of anything less than the design performance specification for the equipment. One of Murphy's Laws is: "All equipment is just waiting to break down." Unfortunately, this is the attitude of most employees. However, in quality assurance programs, employees are taught to expect zero defects. In equipment and facility maintenance, they should be taught to expect zero breakdowns and zero failures. There are six significant causes of poor equipment utilization, as follows:

1. breakdown losses

2. setup and adjustment losses

3. idling and minor stoppages

4. reduced capacity losses

5. quality problems—habitual and occasional

6. startup/restart losses.

Breakdown Losses

There are two major types of breakdown losses that must be recognized: capacity loss and capacity reduction breakdowns. The capacity loss is the easiest to recognize since the equipment ceases to operate. This is what is typically called a breakdown. This is the production stoppage or the stoppage of a service to a facility. The maintenance department responds in an emergency mode and works to quickly restore the operation or service.

The capacity reduction breakdown is more subtle in nature. As the equipment ages, it experiences wear. As the wear continues, the capacity of the equipment begins to decline. Unless careful monitoring occurs, the reduced capacity goes unnoticed or is accepted as normal. In production, this translates into slower operation, lower capacities, and increased labor costs. In facilities, it translates into higher energy and operational costs.

For example, failure to maintain adjustments and calibrations on HVAC systems may result in a 25% increase in energy costs.

While the capacity loss breakdown is the easiest to find and repair, the capacity reduction breakdown represents the largest cost to most corporations. In the majority of cases, the capacity loss breakdown is a technical problem. The capacity reduction breakdown, in most cases, is an organizational problem.

The capacity loss breakdown is caused by the failure of some component on the equipment. Since most preventive/predictive maintenance programs are designed to detect and trend normal wear, other types of wear will cause the majority of these breakdowns. The other types include infant mortality failures, and failures related to poor operating and maintenance activities. If the preventive/predictive maintenance program is effective, most of these breakdowns will be minimal in nature.

The capacity reduction breakdown is generally caused by neglect of a chronic equipment problem. This is a problem that occurs over a long period of time and becomes accepted as a normal fact of operation. A typical example would be equipment that produces a defective product when operating at any rate over 80% of design speed. Instead of taking the time and effort necessary to correct the problem, the organization will issue a memo stating not to run the equipment over 80% of design speed. This results in a 20% reduction in equipment capacity. If this process is repeated over several years, the plant will soon need to invest in new equipment just to meet the necessary production rate. The problem becomes severe because management is focusing on short-term goals and not the long range.

When capacity reduction problems develop, it is more economical to solve the problem than it is to reduce the operating standards. It is merely a matter of examining short-term profits in light of long-term profitability.

The elimination of both capacity reduction and loss breakdowns is included in the zero breakdown concept. If this goal is to be realistic, it takes commitment to a program designed to prevent breakdowns. The program must address the different problems—both organizational and technical—that contribute to the breakdowns. The program will consist of the following 5 basic parts:

maintain basic equipment condition

maintain designed operational standards

restore normal equipment wear

use reliability engineering

eliminate human error.

1) Maintaining basic equipment conditions is a maintenance practice that is ignored in most companies today. When the maintenance group gets occupied with capacity loss breakdowns, trying to keep the equipment operating properly, basic tasks like cleaning, lubricating, adjusting, and tightening are neglected. The resulting deterioration and wear caused by lack of service contribute to additional capacity loss breakdowns, resulting in less service. This snowball effect spells disaster for an organization.

In a true T.P.M. organization, the operators will take responsibility for some of these steps, allowing maintenance personnel to focus on other tasks. The cleaning of the equipment is critical since any contamination will accelerate wear and increase deterioration. For example, contamination on a gearcase shell, usually lubricant covered with dust, develops a barrier between the gearcase and the atmosphere. Since a gearcase shell acts to help dissipate heat developed inside the case, the effectiveness of this design function is reduced. This leads to increased temperature in the gearcase, reducing the effectiveness of the lubricant. The higher temperatures and reduced protection from the lubricant result in accelerated wear and deterioration. This increases the probability of a capacity loss breakdown.

The previous example is just the tip of the iceberg when it comes to the problems that cleaning of the equipment can prevent. Similarly, basic tightening can eliminate problems related to looseness and vibration. Any looseness and vibration accelerates mechanical wear, contributing to deterioration of the equipment. There is a problem related to maintaining proper tightness of mechanical equipment: proper torque adjustments. Just as looseness is a problem, excessive tension applied to fasteners is also an invitation to problems. Excessive tension on a fastener exceeds the designed elasticity of the material. Once these limits are exceeded, as in a bolt, it becomes impossible to maintain tightness. The bolt will continue to stretch, allowing looseness and resulting in mechanical wear. The answer to this type of problem is proper training and standards for tightening and adjustment.

Lubrication is also a key ingredient to maintaining basic equipment conditions. Effective lubrication programs ensure that the right lubricant is applied in the proper quantity to the correct application points. This program also requires training of the operations and maintenance personnel involved. Lack of lubrication accelerates equipment wear. However, excessive lubrication can also damage equipment by increasing temperature levels as the lubricant churns, damaging the equipment and the lubricant. If seals are damaged and leaking, contamination can enter through the defective seal contributing to increased wear of the equipment. If lubricants from different vendors are mixed, without comparing interchange charts, chemical additives will react and raise or lower the lubricant viscosity, the lubricant may thin or coagulate, or the lubricant can become acidic or alkaloid, causing adverse reactions with the very equipment it was supposed to protect. An effective lubrication program will ensure that the operations and maintenance personnel are properly trained in lubrication basics. Regular lubrication points may be color coded indicating what type of lubricant is to be used. There should be charts provided to ensure that the proper amount of lubricant is applied.

The best method to ensure that the proper standards are adhered to in the above areas is to document the maintenance program. Many companies already have this step completed with their preventive maintenance program. However, in order to provide sufficient coverage and detail, the service and inspection instructions should be comprehensive. The instructions should be detailed, since "add lubricant if necessary" is not sufficient. The amount and type should also be included, so mistakes cannot be made. These standards will ensure that the basic equipment conditions are maintained.

2) Maintaining designed operational standards is the component of the breakdown elimination program developed from the original design specifications for the equipment. This means that the equipment should operate at its rated speed and capacity throughout its entire life cycle. The two alternatives are to operate it below capacity or to operate it above its rated capacity. Operating the equipment below its rated capacity is a waste of capital assets. It may cause the company to invest in unnecessary equipment and floor space to produce its product. In a facility it may mean buying extra environmental control equipment (such as HVAC) to maintain the facility. This is an expensive wasteful practice and will certainly impact a company's ability to be cost competitive. Operating the equipment above its rated capacity

will overload it, resulting in excessive wear on the equipment as well as subsequent premature depreciation.

If a company will determine the design specifications for each piece of equipment or system, and operate at these specifications, it will ensure that the maximum value from each asset is obtained.

Once the standards are established, they should be documented and distributed to operational and maintenance personnel. These standards will help them ensure that the equipment is properly maintained and operated. Included in these standards should be the environmental conditions: temperature, humidity, and contamination. Ensuring that the equipment is operated in the specified environment will again help provide the maximum return on investment for each asset. Failure to do so, as previously highlighted, will contribute to excessive wear and premature failure.

Another area of note in the standards for companies that move equipment regularly is construction and installation. Standards would help to ensure that the equipment is properly installed or stored each time it is moved. The standards here could include electrical specifications, proper piping specifications, and mechanical installation specifications. For example, improper physical installation of equipment could allow for excessive vibration, which in addition to causing wear could be a contributing factor to quality problems. Proper attention to maintaining equipment operating standards is a major part of ensuring equipment capacity.

3) Restoring normal equipment wear is a process that continues for the entire life of the equipment. One of the first steps in the process is to set standards for the equipment condition. These standards should be set to a level that will assure equip-

ment capacity. This means a good system of inspection and repair must be set in place.

Engineering calculations such as mean time between failures and mean time to repair should be used to help set repair frequencies. This will provide information necessary to set maintenance replacement and rebuild intervals. These indicators, coupled with good use of predictive maintenance technologies, should prevent any capacity reduction conditions from developing to a level that affects operations.

As the maintenance becomes more controlled, standard repair policies and practices should be developed. These policies and practices are usually in the form of standing or repetitive work orders. These standardized repair policies should include detailed instructions, lists of parts required, and estimated labor resources required. This eliminates the uncertainty and lack of coordination found during many equipment rebuilds. It also will provide the ability to properly schedule the work, with minimum disruptions to the production or operating groups.

A subset of standard repair procedures is the standardization of inventory items and their storage. This means that the repair parts will be consistent and they will be able to be located. Presently, in many companies, the purchasing and inventory groups change suppliers or manufacturers so often that there is continual usage of interchange charts. There is always the situation that parts that were supposed to be interchangeable do not really match. This is usually not discovered until the repair is in progress. This contributes to delays in making the repair and lost production or use of the facility. Standardization of spare parts and spare procedures is one of the most important support areas to ensure restoration and preservation of equipment condition.

4) Reliability engineering is used to correct equipment-related problems that cannot be addressed through normal maintenance or operator-based maintenance procedures. Reliability engineering is generally applied to design weaknesses. Reliability engineering can be used to strengthen parts to extend the life of the equipment. These are usually chronic problems that directly impact equipment capacity. Some of the typical techniques used are wear resistance and corrosion resistance. Elimination of these two problems will contribute major improvements to equipment life cycles.

Another main part of reliability engineering is design or material changes to reduce stress and fatigue of materials. This can be accomplished by more accurate machining of parts, by changing the materials that the parts are made of, or by changing the shape of parts or components. In addition to this, the accuracy with which some components are manufactured can be increased, contributing to less wear on the component and related systems. If used in conjunction with the other capacity assurance techniques, reliability engineering will assure optimum availability.

5) Elimination of human error is an equipment capacity problem with two basic divisions: operations and maintenance. The human error for operations is misuse or abuse of the equipment or facility. The human error for maintenance is lack of maintenance or repair errors.

Operational misuse or abuse of equipment is controlled by training. The cause of the misuse or abuse is first studied. If it is lack of knowledge on the part of the operator, he is properly trained. If it is due to poor operating procedures, these must be updated and distributed. If it is a human–machine interface problem, then the design of the equipment interface must be improved. In some cases, the Japanese use the "poka-yoka" or fool-proofing technique to prevent misoperation of the equipment. This will be some simple device or alteration to the equipment that prevents any mistakes by the operator.

Maintenance errors will either be a failure to perform some maintenance act or an actual repair error. In either case, it is a matter of examining the causes of the problems and correcting the problem. For example, if the problem is a missed maintenance act, such as a Preventive Maintenance (P.M.) service, the cause for missing the P.M. should be investigated. If it was a conflict between the maintenance and operations schedule, then both groups must work toward a resolution, since the corporate assets are affected.

Additional maintenance errors may be corrected by improving working conditions, and by improving the tools and equipment that maintenance utilizes in the repairs. Long-standing chronic problems may be solved by simplifying maintenance repair procedures and troubleshooting processes.

Elimination of the human factor in equipment capacity assurance programs will remove one of the most difficult to detect chronic problems.

Setup and Adjustment Losses

Setup and adjustment losses are incurred when equipment is used to manufacture different products. The definition of setup and adjustment time is the period of time lost when the equipment is finished producing one product and is producing the next product at the specified quality. In some operations, this time period can be considerable. The losses incurred during this time can make a difference between a profit and loss for a company. For example, if setup and adjustment time is reduced, the lot size which is economical to run is reduced. This increases the flexibility of the production scheduling and sales group, allowing them to be more responsive to the customers' requests. The responsiveness to the customer is a measure of the competitiveness of the company; it may make the difference between winning and losing the business.

There are three typical areas in which to make improvements in equipment setups and adjustments.

1) *External Setups:* These activities can take place while the equipment is in operation.

2) *Internal Setups:* These activities can take place only while the equipment is shut down.

3) *Adjustments:* These activities take place while trying to get the equipment to produce a quality product.

Each area can make a major contribution to the overall improvement.

1) *External Setups:* In order to be most effective with external setups, it is necessary to make all assemblies and adjustments before the unit is taken out to the equipment to be installed. This may involve preparation of all tools and equipment that will be required to make the installation. The external part will also ensure that all related tools are ready, such as precision measuring instruments. All measurements that can be set independent of the unit should be made. In some cases, this may involve preheating dies to an operating temperature before installation.

This step can be simplified if the work area is properly organized and equipped. The preparations for each setup should be in a written, step-by-step procedure. This allows all tools and equipment to be in the proper position before the change begins. The three simple reminders are: do not have to search for anything; do not have to move anything; and do not use anything that is not specified in the procedure. If these simple steps are followed, then the rewards will be substantial.

2) *Internal Setups:* The internal setups are performed when the equipment is not operating. This is the actual changing or replacement of the tool, die, etc. The shorter the time it takes to do the internal setup, the faster the equipment can begin to produce a new product. In some cases, the amount of time has been reduced from days to hours and even to minutes. The first step to improving internal setup time is the standardization of work procedures; this enables the crews to be consistent in the change, with no unexpected problems. Most setups require several people to be involved in the process. In order to ensure the effectiveness of the setup, all work should be allocated prior to the actual setup. Personnel should be allowed to work in parallel paths to accomplish as much as possible in a short time period. Some companies will videotape this process and allow the employees to study the tapes, looking for ways to improve the tools or techniques to reduce the actual setup time.

Additional areas to investigate would include the physical equipment itself. Attempts should be made to reduce the number of parts required to clamp the equipment in place. Also, the clamping method should be as simple as possible to allow quick removal and replacement, without compromising the necessary clamping force to produce a quality product. Considerations would include the shapes of the equipment, the weight of the equipment, and the interchangeability of the equipment. If these areas are effectively analyzed, the setup time can be dramatically improved.

3) *Adjustments:* After the equipment has been changed, there is a period of adjustment that is required to ensure that the product meets the quality specifications. In most cases, this time can be considerable. The goal of studying the adjustments is to

find methods that will eliminate all adjustments. If all factors are set to specifications during the internal and external setups, this can be achieved. One area of consideration is the precision of the equipment. Trying to get one more cycle out of a tool or fixture can be more expensive than the replacement would have been. The lost time in trying to make marginal or out-of-tolerance equipment function properly will be considerable. In addition, all of these adjustments will have to be changed when the next setup is made, since the good equipment will require all different settings.

One of the areas where major improvement can be made at low cost is the simplification of the adjustment process. This would include good reference surfaces, simple measurement methods, and simple standard adjustments. In many cases, the go − no go gauges can be used to speed up the process. If the equipment, tools, and procedures are kept simple and effective, the goal of no adjustments can be reached.

Idling and Minor Stoppages

These are areas where the equipment is stopped for a few minutes or is idle due to some upstream operation malfunction in a process or assembly line. These minor stoppages are ignored for the most part, since over a period of time they become accepted as a normal operational characteristic of the equipment. The losses in this area can be considerable, in some cases more than the losses caused by the capacity loss breakdown. The three main types of idling and stoppages are

1. overloads,

2. malfunctions of the equipment, and

3. upstream malfunctions.

The overloading of equipment will trip some type of limiting device. The equipment will shut down till operations or maintenance personnel reset the device. This loss will only be for a few minutes, but if repeated over time, can be greater than anyone realizes. For example, if a piece of equipment experiences a 10 minute stoppage per shift, on a 3 shift per day operation, running 5 days per week, the monthly total of downtime for the equipment is 10 hours. If you consider what an hour of downtime costs on a piece of equipment, the true cost of this type of loss can easily be in the thousands, tens of thousands, or even hundreds of thousands of dollars. These costs multiply rapidly if the equipment is not standalone, but part of a production process. The delay on the first process will contribute to similar delays on the processes downstream. In a Just-in-Time or C.I.M. environment, the total loss may be tremendous. Yet these small losses go unnoticed and uncorrected by most companies. In order to be cost effective, these losses must be eliminated.

The steps necessary to correct these minor stoppages are basic maintenance practices. Each delay should be investigated. The root/cause analysis should be performed, with a fishbone diagram employed to track the common problems. The common problems should be analyzed first, with the more complex problems being addressed after these are eliminated.

The proper basic maintenance will also help to eliminate these problems. In many cases, keeping the equipment in its proper repair and adjustment, ensuring that it is close to its optimum position in its life cycle, will help to eliminate most idling and minor stoppages.

One key point should be remembered: act on each problem. Simple neglected problems can develop into serious complex problems.

Reduced Capacity Losses

The reduced capacity losses are defined as the difference between the rated capacity of the equipment and the capacity at which it is operating. The two largest areas that impact the capacity of the equipment are the speed and volume of the output.

The first problem is that the capacity of the equipment is unknown. This is true particularly in older equipment. The true speed and output of the equipment should be identified. This should become the new capacity rating of the equipment. Then, reaching the capacity will usually entail the correction of many minor problems and defects that prevent the equipment from reaching this level. As with many other problems previously discussed, operations personnel learn to accept less than design capacity from their equipment over long periods of time.

The second problem involves the failure of operations and maintenance to correct problems as they are uncovered. Allowing the equipment to limp to the next shutdown or outage may cost more than taking it off line and correcting the problem. The lost capacity, coupled with the increased costs for operations and maintenance, will be greater than continuing with the reduced production output, even on a temporary basis.

The rule to remember in this area is to not accept anything less than capacity rated performance from any equipment. If this rule is followed, all necessary steps will be taken to correct minor problems before they become major problems.

Quality Problems—Habitual and Occasional

Maintenance-related quality problems can be divided into two broad classifications: habitual and occasional. The habitual defects are caused by equipment operating with some deterioration that has come to be accepted as normal operation. The occasional defect is caused by some apparent malfunction or problem that has suddenly developed.

The habitual defects are caused by equipment-related problems. Again, these are problems that have developed over a longer time period and are neglected. It may be a case of short-term fixes or the "band-aid" approach to solving the problem. However, the true problem is never addressed. In order to compensate for the problem, other adjustments are made. These adjustments affect other systems on the equipment, and soon the operational standards are affected. This will have varying effects on quality. Only by restoring the equipment to its original operating standards will the habitual defects and their quality-related problems be eliminated.

The occasional defect is caused by a failure in one of the systems or parts of the equipment. The problem is easily discernible. The cause and corrective action are quickly and easily discerned. The action to remedy the problem should be taken immediately. In no case should an occasional defect be neglected; it should be addressed at once.

In studies conducted at some plants, over 50% of all quality problems are maintenance related. Failure to correct both habitual and occasional equipment problems can lead to a very expensive quality-related problem.

Startup/Restart Losses

These losses are incurred each time a process must be shut down and restarted. These losses may involve producing an unacceptable product while the equipment reaches a certain operations parameter such as temperature or speed.

These losses may be considerable, so any equipment shutdowns related to a failure must be avoided. These losses are often overlooked when considering downtime costs. So instead of the costs being just the lost production while the equipment was shutdown, they may also include lost production while shutdown and startup were occurring. In some plants, this may involve hours of lost production per occurrence.

By applying many of the techniques mentioned previously, these losses can be kept to a minimum. In no instance should these losses fail to be included in the total lost production cost calculated during a function loss or function reduction breakdown.

Other Related Costs

There are other costs, beyond the aforementioned six significant losses, that have not been specifically mentioned. These losses include the manpower loss, the manpower used at a premium cost to make up lost production, and energy losses. These losses are self-explanatory, and the solutions have either been discussed or are self-evident.

In conclusion, it should be noted that these costs combined can make up 30% to 50% of the total production costs. From a facility standpoint, the loss can be measured in poor worker productivity, poor appearance and image to clients, and lack of community support for the company.

Measuring the Equipment Performance

Measuring equipment effectiveness must go beyond just the availability or the uptime. It must factor in all issues related to equipment performance. The formula for equipment effectiveness must include the availability, the rate of performance, and the quality rate. This allows all departments to be involved in determining equipment effectiveness. The formula could be expressed as

$$\text{availability} \times \text{performance rate} \times \text{quality rate} = \text{equipment effectiveness.}$$

The availability equals the required availability minus the downtime, divided by the required availability. Expressed as a formula, this would be

$$\frac{\text{required availability} - \text{downtime}}{\text{required availability}} = \text{availability.}$$

The required availability is the time the equipment is in production minus the miscellaneous planned downtime, such as breaks, schedule lapses, meetings, etc. The downtime is the actual time the equipment is down for repairs. This is also sometimes called breakdown downtime. The calculation gives the true availability of the equipment. It is this number that should be used in the effectiveness formula. The goal for most Japanese companies is a number greater than 90%.

The performance rate is the ideal or design cycle time to produce the product, multiplied by the output, divided by the operating time. This will give a performance rate percentage. The formula will be

$$\frac{\text{design cycle time} \times \text{output}}{\text{operating time}} = \text{performance rate.}$$

The design cycle time (or production output) will be in some unit of production, such as parts per hour. The output will be the total output for the given time period. The operating time will be the availability from the previous formula. The result will be a percentage of performance. This formula is useful for spotting capacity reduction breakdowns. The goal for most Japanese companies is a number greater than 95%.

The quality rate is the production input into the process or equipment, minus the volume or number of quality defects, divided by the production input. The formula is

$$\frac{\text{product input } - \text{ quality defects}}{\text{product input}} = \text{quality rate.}$$

The product input is the unit of product being fed into the process or production cycle. The quality defect is the amount of product that is below quality standards—not rejected (and there is a difference)—after the process or production cycle is finished. The formula is useful for spotting production quality problems, even when the poor quality product is accepted by the customer. The goal for the Japanese-based companies is a number higher than 99%.

Combining the total for the Japanese goals, we find that

$$90\% \times 95\% \times 99\% = 85\%.$$

To be able to compete for the national T.P.M. prize in Japan, the equipment effectiveness must be greater than 85%. It is sad to say that the equipment effectiveness scores in most U.S. companies barely break 50%. It is little wonder that there is so much room for improvement in the typical equipment maintenance management program.

3

Pre-T.P.M. Activities

Pre-T.P.M. activities are disciplines that must be put into place prior to beginning a T.P.M. program. Most of these activities are designed to give the organization the necessary information, structure, and discipline to be successful in implementing T.P.M. These activities would include

developing the maintenance organization

equipment information management

inventory information management

organizational assessment.

Developing the Maintenance Organization

The development of the maintenance organization is an important step before beginning a T.P.M. program. It is a step that many organizations, caught up in the spirit of starting a new program, often neglect. However, since most of the equipment information and operator training comes from the maintenance organization, it is only logical to ensure that it is properly functioning before focusing on the operations or production departments.

The development of the maintenance organization is important in the following four main areas:

a. the control system

b. information management

c. the organizational structure

d. staffing.

The proper development of the maintenance organization must first focus on the goals and objectives for the maintenance department. It is like building a bridge; you may know where you want to go, but without knowing where you are, it is difficult to start. The objectives for the maintenance department may be basic, becoming more advanced as the organization matures. For example, the following may be typical objectives for the maintenance department.

Determine the maximum production (or condition of the facility) at the lowest cost, while producing the highest quality and at the optimum safety standards.

Identify and implement cost reductions.

Provide accurate equipment/facility maintenance records.

Gather and record all necessary maintenance cost information.

Optimize all maintenance resources.

These objectives will vary from plant to plant or facility to facility, but clear objectives should be established for each maintenance organization. The above objectives should be used only to help each organization focus on its own objectives. It is difficult to state globally what the objectives must be, since maintenance is a support function and must tailor its objectives to support and complement the goals and objectives of the rest of the corporation.

Once the objectives have been set, it is necessary to set a control system in place to monitor and report on the progress made in achieving these objectives. A typical control system is shown in Fig. 3-1. This control system is designed to find exceptions to the stated objectives and supporting strategic programs. For example, a supporting strategy to identify and implement cost reductions might be the control of maintenance overtime. A good goal is 5% of total time worked. If the control system is used, there is no need to take any action when the overtime is 5% or less. If the overtime exceeds 5%, then an exception report should be generated showing that one of the indicators has been exceeded. The true problem should be discovered and a corrective action implemented. The overtime indicator should then be monitored to be sure the corrective action actually solved the problem.

The enigma in this case is that maintenance problems typically have multiple causes. It might be illustrated as a doctor diagnosing a disease. Many diseases have similar or identical symptoms. Only by running various tests and using different techniques to actually determine the problem can a doctor make a proper diagnosis. He can then treat the true disease quickly and effectively.

In maintenance organizations, the same problem exists. Instead of solving the true problem, the symptom is treated. For

MAINTENANCE MANAGEMENT CONTROL SYSTEM

* Establish Goals, Objectives, Policies, and Procedures

* Establish Permissible Variance from the Guidelines

* Measure the Performance and Compare to the Guidelines

* Compare the Evaluation to the Permissible Variance

* Identify the Exceptions to Tolerance

* Determine the Cause for the Exception

* Determine the Corrective Action

* Plan the Implementation of the Corrective Action

* Schedule the Implementation of the Corrective Action

* Implement the Corrective Action

* Evaluate to Avoid Overcompensation or Failure of Action

Figure 3-1. A sample maintenance management control system.

example, if maintenance costs increase, the first action is to reduce headcount or inventory levels. It may be that these will appear to solve the problem, but the disease still exists. Once the organization stabilizes after the reduction, the disease will surface again. It is only by using techniques requiring multiple indicators that the true problem can be uncovered. Consider the typical problems listed in Fig. 3-2. How many of these problems are related? How many have the same symptoms? There are even cases where two or more of these problems can be interrelated. Only by having good, reliable data will the true problem ever be uncovered and solved.

How does the organization decide on what information to track, and how does the organization gather that information? It is a matter of understanding the corporate goals and objectives, and then structuring the maintenance goals and objectives to support them. Once the maintenance goals and objectives are established, then the data required to monitor the results are identified. This is usually done by a maintenance indicator. The data necessary to support the indicator are identified, and it is then determined how to collect the data. Some data will come from operations or facilities or even the financial department. However, the majority of the data necessary to support the indicators must come from the maintenance department.

Highlighting this point is the typical problem discovered by the financial officers: "maintenance costs are up this month." The next question is "why?" The financial department may say we

COMMON MAINTENANCE EFFICIENCY PROBLEMS

* Equating Maintenance and Production Staffing Levels

* Crew Location Problems: Area–Central–Combination

* The Effect Reactive Maintenance Has on Productivity

* High Levels of Unexpected Downtime and the High Cost

* Random and Seasonal Fluctuations

* Poor Communications: Maintenance–Operations–Management

* Lack of Equipment Standards

* Lack of Qualified Personnel

Figure 3-2. Common maintenance efficiency problems.

used more materials than normal. The next question is "why?" The financial department may be able to determine what cost center or account code the parts were used in, but only the maintenance department knows what job they went to or what piece of equipment they were used on. How does the maintenance department track this information? It is through the use of a comprehensive and disciplined work order system.

Work Order Systems

The maintenance work order system is analogous to the production order system. In the production order system, an order number is used to track a customer's request from the initiation in sales through the entire plant and out to the customer's site through shipping. The production order is filed so that any questions or problems with the customer's orders can be resolved quickly and correctly. It is also valuable because when the customer places another order for the exact same items, the order is pulled from history and duplicated.

The maintenance work order system is almost identical. The customer (operations or facilities) places the order for a repair or service. The maintenance department assigns the work an order number. Then all activities related to the completion of the work order are recorded on the work order. When the work order is completed, all costs are posted to it and it is filed in history. If the request is ever placed again, then the history can be reviewed and the order completed as before.

Just as reports and analyses are conducted on the production system to ensure the profitability of the company, the same is done to the maintenance information. This ensures that complete, cost-effective results have been achieved. Any problems or questions are easily referenced and resolved to the customer's satisfaction.

This scenario should help everyone to appreciate that maintenance not only can be, but needs to be, managed and controlled like other parts of any business. The work order is the information gathering device used to feed data to the maintenance control system. The typical objectives for work order systems are listed in Fig. 3-3.

While the information necessary to accomplish these objectives may vary from site to plant to facility, the information gathered on the work order should always be focused on supporting the information needs. Miscellaneous information that does not contribute to satisfying the needs only complicates the information gathering process. In most cases too much information is just as frustrating as too little information. This is why the information needs and the control system indicators must be identified early in the process.

Information Management

After the control system is set in place and the work order system is producing information, the next step is to manage the information. This was the problem in the early days of comput-

WORK ORDER OBJECTIVES

* A Method for Requesting, Assigning, and Following Up Work

* A Method of Transmitting Job Instructions

* A Method for Estimating and Accumulating Maintenance Costs

* A Method for Collecting the Data Necessary for Producing Management Reports

Figure 3-3. Examples of objectives for work order programs.

erized maintenance management systems; they gathered a lot of data, but no one knew what to do with them. This is another reason why all the information gathered should have a purpose. If the indicators are identified during implementation of a program, the tendency to gather too much data is eliminated.

The information gathered should be identified as to the timeliness of the reports that will be created. The typical distribution for reports is on a daily, weekly, monthly, quarterly, and annual basis. When creating maintenance reporting requirements, the previously mentioned point must be reinforced; exception reports are used to manage, summary reports are used to review trends, and detailed reports are used to archive data and review specifics.

Each of these reports has its use. But under no circumstance should any manager have all types of reports on his desk each day. Being buried in reports, not having the time to understand the data you are given, is just as bad as not being given any data at all. In either case, the manager will not have the complete picture.

One additional point on the reports regards the accuracy of the data. These reports will only benefit the company if a disciplined approach is taken to the work order system. If repairs are reported as preventive maintenance tasks, if parts are charged to the wrong work orders, and if emergencies are reported as normal repairs, then the true picture will always be distorted. The analysis is only as good as the data. Dedication to the basics—such as gathering accurate information—will ultimately spell success or failure of any maintenance improvement program.[1]

---

[1] For a complete description of reports that may be used to manage maintenance, see the book *World Class Maintenance Management* (New York: Industrial Press, 1990).

Maintenance Organizations

The structure of the maintenance organization is important to the effectiveness of the maintenance service it provides. Three common types of organizational structures are: centralized, area, and combination. There are advantages and uses for each type of structure.

The centralized organization is generally found in companies that are built around a facility or plant that is confined geographically. Maintenance workers have very little travel and can reach all equipment in a relatively short period of time. This arrangement is usually supplemented by a centralized maintenance stores arrangement. This allows for lower maintenance inventories and good service from the stores counter.

Area organizations are used where the equipment is spread out over a larger area and travel time becomes a problem for the maintenance personnel. The communication and logistics problems increase in an area organization. There is an important benefit to area organization in that it tends to build team spirit and cooperation between the area maintenance worker and the operations personnel. This is an important step to the beginning of a T.P.M. program. The maintenance people also begin to develop ownership in the equipment—another plus in the T.P.M. team development. A problem that may develop in this organization is the concept of area stores. This increases inventory levels and requires additional storage areas. However, as maintenance becomes more proactive, and as planning and scheduling become more disciplined, a delivery system can be set up that will reduce the need for too many area stores.

The combination organization allows the flexibility of the centralized organization and still gains the benefit of the area organization. The team work and ownership of the area organization are still fostered, but the support from the central organization is available when needed. This organizational arrangement is the last step maintenance will take before entering a team-based organization. The maintenance team member will coordinate all maintenance activities for his team, utilizing the centralized manpower as is required to complement his area efforts.

The combination organization is the goal that each larger company should work toward to set the foundation for T.P.M. The smaller company may still stay centralized, but could assign the preventive maintenance and repair responsibility to certain individuals. In one respect, they become "quasi-area" assignments. This still enables the ownership and teamwork concepts to begin to develop.

Maintenance Staffing

The staffing of the maintenance organization should be divided into two separate groups: craftworkers and support personnel.

The craftworker's staffing is determined strictly by the backlog of work for each craft. In a multiskilled environment, this may be the total maintenance workload. The craft backlog is the amount of work that is documented as needing to be performed by the craft. The work should be documented on the work order form. The work that is counted in the backlog is only the work that is ready to schedule or that can be performed at the present time. This removes work that is waiting on:

engineering support

spare parts

approval

shutdowns, outages, or rebuilds.

The formula for calculating the backlog is as follows:

$$\text{backlog (in weeks)} = \frac{\text{total planned hours ready to schedule}}{\text{true craft capacity}}$$

The true craft capacity is the total hours scheduled for the craft for a week minus schedule interruptions. Schedule interruptions should include average hours spent on emergencies, absenteeism, vacations, and routine or (in some cases) preventive maintenance work. This leaves the total hours that the craft will actually deduct from the backlog. An example of this process is illustrated below.

Total employee hours scheduled for next week (10 men × 40 hours)	400 hours
Total overtime to be worked (average for last 3 months)	40 hours
Total contractor labor (2 men × 40 hours)	80 hours
Gross labor hours available	520 hours
Average emergency work (50% for the last 3 months)	260 hours
Average absenteeism/week	10 hours
Average vacation hours/week	10 hours
Average routine (nonbacklog) hours/week	40 hours
Total deductions	320 hours

Gross minus deductions = 200 hours.

These 200 hours represent what can realistically be expected to be completed from backlog work for the week. This is the number that should be used to determine the true backlog. The calculation would be as follows (assume 2000 hours in the backlog):

$$\text{Backlog in weeks} = \begin{array}{cc} \text{Example 1} & \text{Example 2} \\ \dfrac{2000 \text{ hours}}{520 \text{ hours}} & \dfrac{2000 \text{ hours}}{200 \text{ hours}} \end{array}$$

In the first example, the craft backlog would be 3.8 weeks of backlog. But it would be impossible to complete the work in the backlog given the constraints placed on the craft time available.

In the second example, the craft backlog is 10 weeks, which is a realistic time period in which to complete the work. Normal maintenance backlogs are optimized when they are kept in the 2–4 week range. Some companies will allow a 2–8 week backlog for a craft. The greater the number of weeks in the backlog, the longer the requesting department has to wait on their work request. If the backlog gets too large, the temptation is to request the work on an emergency basis, circumventing the planning and scheduling process, and thereby increasing the cost to do the work.

If the backlogs are kept in the 2–4 week range, the maintenance disciplines are easier to maintain. If backlogs begin to increase, overtime and contract labor may be used to see if the increase is temporary. If the increase is a permanent trend, new craftworkers may have to be hired. If the backlog decreases within two weeks, the overtime should be eliminated, and perhaps contract labor decreased or eliminated. If the backlog continues to decline, then the staffing may need to be examined and employees repositioned. Under no circumstances should maintenance labor be increased or decreased without accurate backlog figures. Changes made independently of the backlog are arbitrary and dangerous to the effectiveness of the maintenance organization and ultimately to the corporation.

Staff Personnel

Proper staff support levels is one of the most debated topics in maintenance. The correct method of determining the staff levels begins with the correct determination of the craft staffing. Once the total number of craftworkers is determined, the correct level of first line supervisors and planners can be determined. The normal ratios are 1 planner for every 15–20 craftworkers, and 1 supervisor for every 10 craftworkers. If we considered an organization of 40 craftworkers, it would translate into

4 supervisors

2 planners

1 manager.

The number of clerks is determined by the amount of data the organization is required to keep to meet its information management objectives. Care must be exercised not to overstaff in the area of clerks. In some cases, the number of clerks increases because the planners treat them as "their secretaries." Planners will handle most of their own paperwork. The clerks are supposed to help manage the information flow, such as reporting, filing, timekeeping, etc. In the example above, the clerical load, if everyone is doing his or her job, may be 1 or at the most 2.

As a rule of thumb, staff levels for a maintenance organization should never exceed 25% of the craft workforce (this assumes that the number of craftworkers is correct). It must be kept in mind that the level of clerical staff support is ultimately determined by the amount of data the organization is required to keep. If a company requires maintenance to perform asset tracking, then the clerical function will be staffed at a higher level since the data tracking requirements are much higher. Geographic layout of the plant, skill levels, sophistication of the equipment, and other factors prevent any firm rules for staffing to be specified. Each plant, even within the same company, should be studied so the correct staff levels can be determined.

Equipment Information Management

Since a T.P.M. program is designed to increase equipment effectiveness, it is important to identify the particular asset or equipment that is to be maintained. The equipment information management pre-T.P.M. activities are

   a. equipment identification

   b. nameplate information

   c. equipment history.

Equipment Identification

Equipment identification starts with deciding the difference between a piece of equipment and a component or part. In some companies, a motor is a piece of equipment; in other companies, it is a part or component. Why is this a problem? Each piece of equipment will be required to have a detailed history, made up of the completed work orders for the piece of equipment. While the parts or components will require some data to be kept, it is not the volume of information that the equipment requires. In making the decision, an equipment item should be equipment where a detailed historical repair record is required. The repair cost, frequency, date of repair, and year-to-date and life-to-date costs are all common data that are kept. The more equipment that is listed, the greater the clerical load. Some companies set a dollar limit or size limit on various components that will be classified as equipment.

Once the items have been classified as equipment, it is necessary to give them a specific equipment identifier. This is usually called the equipment number. It is important to use an intelligent equipment numbering scheme, otherwise it will be difficult, later in the program, to track the information required to improve equipment effectiveness.

It is common to see sequential numbering schemes used to identify the equipment. While for just tracking data, these may seem to work, they fail to provide the flexibility that is required for reporting. The following example will provide some details on numbering schemes and their usefulness.

When numbering equipment, it is wise to consider the layout of the plant or facility. The first set of digits could be used to indicate the area (see Fig. 3–4A). The second series of digits could be used to identify a specific unit within that area (see Fig. 3–4B). The third series of digits could be used to indicate the

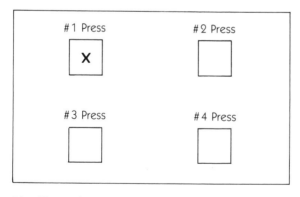

—PLANT—

1	2	3
4	5	6

Area = 1–6

Area = 006 or 06

A

#1 Press ☒	#2 Press ☐
#3 Press ☐	#4 Press ☐

Identifier = 1–4

Identifier = 001

B

Figure 3-4. Areas of a facility and specific equipment in an area can be assigned individual number codes for easy identification.

type of equipment to which the number applies (see Fig. 3–5A). The last set of digits would signify the particular equipment item that is being identified (see Fig. 3–5B). For example, when the whole series is put together, the number would be:

006-001-001-003.

It is the number 3 hydraulic system on unit 1 in area 6. While this seems like a lot of work, there is a reason for it all. In a manual system, the numbers will not be useful other than to identify the specific equipment. But in a computerized system, it allows tremendous flexibility for reporting. For example, if you wanted to find all the breakdowns on hydraulic systems in the plant, you know that the third series of numbers is the equipment type, and the identifier for hydraulic systems is 001. It is simple to write a

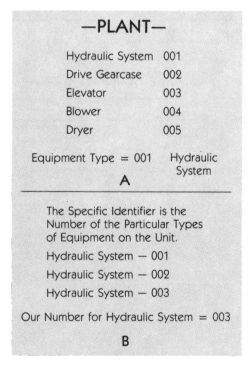

Figure 3-5. Additional codes can be used to identify the type of equipment and particular functions.

query for the computer to find all repair records for type 001. This flexibility is lost when sequential numbering schemes are used.

A warning on numbering schemes: keep them simple. If they are long (over 12–15 digits) and are unintelligent, they are difficult to remember and will quickly fall into disuse. The best numbering schemes are short and intelligent.

Nameplate Information

After the equipment has been identified, it is necessary to gather information about each individual equipment item. This information is about the components that make up the equipment. Continuing the example of the hydraulic system, the following would be some of the nameplate information:

Motor ID #	2884-3848
RPM	1800
Frame	32-T
Voltage	440 VAC
Shaft OD	1⅞″
Bearing Type	Fafnir 803
Pump ID #	234-341
RPM	1800
GPM	200
Frame	486-D
Manufacturer	Vickers

While this is just a sample of the data required, there is much more that can be kept. It is up to the managers to decide what information is going to be required to meet the goals and objectives of the organization. When making this decision, managers should remember that it is easier to have the information recorded and stored than it is to have to go out to the physical site to look at the information each time a question arises.

It is time consuming to gather this information. It is estimated by consulting firms that it takes 1 hour per record to gather the data and enter them into a computer. This step will save many times more than that in the future, when the maintenance technicians do not have to go to the equipment for this information each time they need it.

In addition to the nameplate information, it is necessary to keep drawings current. This is a benefit to maintenance technicians while troubleshooting and effecting repairs on equipment. Also, the vendor's manuals and bulletins should be kept current; this will keep the delays waiting for information at a minimum.

Equipment History

The equipment history is a log of repairs kept for each equipment item. This allows for repair records to be utilized in helping to determine preventive maintenance frequencies, overhaul intervals, and replacement cycles. Some of the information that needs to be kept for the equipment history includes

the type of work

the crafts involved

the date of the work

the dollar value of the repair

any specific detail information.

This information needs to be compiled into a useful format that can be quickly analyzed by the managers, supervisors, or craftworkers. This is where the computer is a big aid to processing the data. Summary reports and exception reports are most useful to find equipment that needs analysis or repair.

The equipment history is the true goal for the work order

system. The equipment history is the database from which most decisions will be made. The validity of the decisions will depend completely on the accuracy of the data that are gathered. If work order history is available at the start of the T.P.M., the accuracy of the information is vital. If the information is not reliable or accurate, it should not be added to the history. Future parts of the T.P.M. program will be based on this information. Inaccurate or unreliable information will damage the future of the program.

Inventory Information Management

Inventory information management is similar to equipment information management. It is a matter of gaining control of the information about maintenance spares and inventory items. This is performed in the same basic three steps as the equipment information as follows:

part identification

part information

part history.

Part Identification

This is the process of numbering all the inventory items with a specific scheme for identification. Some companies will use the vendor's or manufacturer's part number or part identification. This is not a good idea. From time to time, companies will change vendors, manufacturers will change part numbers, or items will be discontinued. Since the part number is a key number that all related information is tracked to, it must *never* be changed. Particularly with computerized inventory systems, where the part number is a record key, changing the number results in loss of history, detail information, and equipment cross references. It is recommended that each company develops its own part numbering schemes. A sample might look like:

030-005-0235
(A) (B) (C)

The "A" grouping is the particular classification of the item. It might be electrical, electronic, power transmission, hydraulic, pneumatic, etc. The "B" grouping is the type within each classification. For example, the power transmission group might have types such as roller chain, V-belts, flat belts, gear drives, etc. The "C" grouping is the specific part.

Evaluating the example above, it could be a V-belt. The translation would be as follows:

030	=	power transmission
005	=	V-belts
0235	=	the specific belt 4L-235.

Each company would have to evaluate what numbering scheme it would develop; but again, as with the equipment, it should be an intelligent numbering scheme.

Part Information

The information required to be kept for each part includes the following:

locations

quantities

details

manufacturers

vendors

order information.

This information is very similar to the detailed information that is kept for the equipment. One major difference is specifying each location or locations where the parts are stored. The locations should be specific so when the parts are needed, they can be quickly and easily found.

If an inventory storage area is unorganized, it will take approximately 1 hour per item to find and record the detail information.

Part History

The part history is the usage of the part for the last 12–24 months. The usage patterns that are recorded can help set the ordering patterns, stocking levels, and purchasing quantities. This information can be gathered from purchasing if it is not presently kept by the stores group.

The part history should not be used unless it is accurate. If the information is unreliable, then all the decisions involving parts will be unreliable.

Organizational Assessments

By the time a company has gathered the above-mentioned information, it will have detailed information about its plant's or facility's present condition. But the information is still fragmented. The next step is to pull the information together in an organizational assessment. The organizational assessment used to be called a maintenance audit, but that is really a misnomer. Since maintenance is a service organization, many of the attitudes, ideas, and methodologies are influenced by other parts of the organization. Any influencing factors must be taken into consideration while assessing the condition of the maintenance organization. It is truly an examination of the attitudes of the

The Maintenance Organizational Maturity Grid

Measurement Category	Stage 1 Uncertainty	Stage 2 Awakening	Stage 3 Enlightenment	Stage 4 Wisdom	Stage 5 Certainty
Corporate/Plant Management Attitude	No comprehension of maintenance prevention; fix it when it's broken [1]	Recognizes that maintenance could be improved, but is unwilling to fund [2]	Learns more about R.O.I.; becomes more interested and supportive [3]	Participative attitude; recognizes management support is mandatory [4]	Includes maintenance as a part of the total company system [5]
Maintenance Organization Status	REACTIVE: Works on equipment when it fails; otherwise very little productivity [1]	CONSCIOUS: Still reactive but rebuilds major components and has spares available when failures occur [2]	PREVENTIVE: Uses routine inspections, lubrication, adjustments, and minor service to improve equipment M.T.B.F. [3]	PREDICTIVE: Utilizes techniques such as vibration analysis, thermography, spectrography, N.D.T., sonics, etc., to monitor equipment condition, allowing for proactive replacement and problem solving instead of failures [4]	PRODUCTIVE: Combines prior techniques with operator involvement to free maintenance technicians to concentrate on repair data analysis and major maintenance activities [5]
Percentage (%) of Maintenance Resources Wasted	30 + % [1]	20–30% [2]	10–20% [3]	5–10% [4]	Less than 5% [5]
Maintenance Problem Solving	Problems fought as they are discovered [1]	Short-range fixes are provided; elementary failure analysis begins [2]	Problems solved by input from maintenance, operations, and engineering [3]	Problems are anticipated; strong team problem-solving disciplines are utilized [4]	Problems are prevented [5]

	[1]	[2]	[3]	[4]	[5]
Maintenance Workers, Qualification and Training	Poor work quality accepted; rigid craft lines; skills outdated, skills training viewed as unnecessary expense; time in grade pay; low worker turnover/apathy	Workers' lack of skills linked to breakdowns; trade/craft lines questioned; skills obsolescence identified; training needs recognized; traditional pay questioned	Quality + Quality = Quality; expanded/shared job roles; a few "critical skills" developed; training expenses reimbursed; new pay level for targeted skills; increased turnover/fear of change	Quality work expected; "multiskill" job roles; skills up to date and tracked; training required and provided; pay for competency progression	Pride and professionalism permeate; work assignment flexibility; skilled for future needs: operators trained by maintenance, ongoing training; percent of pay based on plant productivity; low employee turnover/high enthusiasm
Maintenance Information and Improvement Actions	Maintenance tries to keep records, disciplines are not enforced, poor data	A manual or computerized work order system is used by maintenance; little or no planning and scheduling	A manual or computerized work order system is used by maintenance, operations, engineering; planners used; scheduling enforced	A computerized maintenance control system is used by all parts of the company; information is reliable and accurate	A maintenance information system is integrated into the corporate operation
Summation of Company Maintenance Position	"We don't know why the equipment breaks down; that is what we pay maintenance for. Sure, our scrap rates are high, but that's not a maintenance problem."	"Do our competitors have these kinds of problems with their equipment? Scrap is costing us a bundle!"	"With the new commitment from management, we can begin to identify and solve problems."	"Everyone is committed to quality maintenance as a routine part of our operational philosophy. We can't make quality products with poorly maintained equipment."	"We don't expect breakdowns and are surprised when they occur; maintenance contributes to the bottom line!"

Figure 3-6. The maintenance organizational maturity grid.

company toward maintenance. In order to assist companies in assessing their maintenance attitudes, the maintenance management maturity grid was developed. This grid is patterned after the "Quality Management Maturity Grid" published by Phillip Crosby in the book *Quality is Free*. The grid is pictured in Fig. 3-6.

In addition to the grid, some of the following indicators should be used to help benchmark the maintenance organization.

Indicator	Goal
Supervisor-to-Craftworker Ratio	1 to 10–15
Planner-to-Craftworker Ratio	1 to 15–20
Labor Productivity	Greater than 60%
Planning Efficiency	
Labor	Greater than 90%
Materials	Greater than 90%
Scheduling Efficiency (Weekly)	
Labor	Greater than 90%
Materials	Greater than 90%
Percentage of Overtime	Less than 5%
Size of Craft Backlogs	2–4 weeks
Service Level of the Stores	Greater than 95%
Equipment Availability	Greater than 95%
Percentage of Maintenance Work Covered by Work Orders	Must be 100%
Percentage of Planned Man Hours Compared to Total Man Hours	Greater than 90%
Preventive Maintenance Cost Compared to Total Maintenance Costs	Must be at Least 30%
Labor Costs Compared to Material Cost	Should be 50%:50%
Number of Maintenance Employees Compared to Plant Population	15%–25%
Total Maintenance Costs Compared to the Cost of Goods	Varies, 4%–8%
Maintenance Costs per Unit Produced	Site Specific

Maintenance Costs per Square Foot Maintained	Varies by Site
Production Loss Caused by Maintenance	Varies by Site
Estimated Replacement Value per Maintenance Worker	Varies by Site
Union Attitudes	Varies by Site
Management Support	Varies by Site
Corporate Environment	Varies by Site

Once these base indicators, or the utilization of other indicators, have been established, the plan to develop the maintenance organization can be formulated. The present benchmark should be documented and the projected improvements established.

Having the facts and projections makes the next step—the presentation of the plan—easier to prepare. The evaluation for benchmarks could be compared to building a bridge. You may know where you want the bridge to end, but you must know where it is to start. Developing the plan to implement T.P.M. requires understanding where you are starting. This is one of the most important steps before beginning the T.P.M. program.

II Developing
the Implementation Plan

In Part I of this book, we discussed the theories and philosophies of Total Productive Maintenance. We also discussed the benchmarking basic maintenance indicators. The vision of T.P.M. for a particular site must start with these steps. First, it is hard to develop a T.P.M. plan if you do not understand what it is. Second, you cannot set this as a goal unless you know how far away from it you really are.

In the second part of this book, we discuss the steps necessary to implement T.P.M. Each company can use these steps to design its own implementation plan. Most of the steps can be modified or altered to meet the needs of each site. All of the steps can be rearranged, if necessary, to suit the priority of each company. Chapter 4 addresses the exception: gaining management support and then keeping it is the highest priority for any organization. If all the reasons for failures of maintenance improvement programs, including T.P.M., were compared, lack of management support would lead the list by far.

The ten-step plan presented here is designed to work for U.S. companies. The approach is different than the Japanese implementation approach. The management style of U.S. companies, their reporting requirements, the financial restraints, and the quarterly profit reporting, all require a different adaptation of techniques and technology. The ten-step plan was shown in Fig. 1–9. The rest of the book presents a detailed explanation of each of these steps.

4

Developing the Long-Range Plan and Convincing the Corporation

The development of the long-range plan translates into looking at the corporate strategic plan and developing a maintenance plan to support it. Once the analysis of the organization has been completed (cf. Chapter 3), the benchmark is set for plan development. By careful analysis of the optimum ranges for some of the indicators, areas requiring improvement can be identified. However, the primary consideration should be the equipment effectiveness formulas. All of the maintenance indicators should be used to help find problems when the equipment effectiveness is low in particular areas.

The highest equipment effectiveness possible is the primary goal for all organizations within a company, and this includes maintenance. So all of the indicators that are used to measure performance should ultimately be related back to the equipment. For example, labor productivity is one indicator that most companies are concerned with. How are the two related? If the labor productivity is low, it indicates a lack of planning and scheduling. This is usually because the level of emergency work is high. High levels of emergency work will be due to a large number of breakdowns (either capacity loss or capacity reduction). This is one of the three main components of the equipment effectiveness formula. So the indicator (labor productivity) will help identify the problem area.

The selection of the indicators that will be used to measure maintenance performance should be related back to the equipment effectiveness. This prevents the selection of arbitrary indicators over which maintenance has no control. In determining the long-range plan, consider the following example.

Area of Consideration	Present	Goal
Organization	Centralized	Area–Team
Supervisors	1:5	1:10
Planners	None	1:20
Employee Training Dollars	$200/Employee	$1000/Employee
Supervisor Training Dollars	$500/Supervisor	$1200/Supervisor
Work Order Utilization	50%	100%
Equipment Effectiveness	45%	90%
Etc.		

This is a simple format of the indicators and setting goals and objectives for the long-range plan. The maintenance indicators selected must be tied back to the equipment effectiveness, otherwise we continue the "ivory tower" and "turf protection" that are so common in companies today.

The next step in developing the plan is to translate it. This translation is not to another national language, but to a business language: Financial. The terms labor productivity, availability, and equipment effectiveness are vague and many times misunderstood by upper management. The translation of the maintenance terms to business terms involves the preparation of a cost justification. The following section is an example for those unfamiliar with the process.

System Cost Justification

Equipment Downtime Costs

1. Annual sales per year (total)   _____

2. Annual production labor costs per year   _____

3. Potential savings due to lost sales and production costs per year (1 + 2)   _____

4. Percentage of maintenance downtime per year (try to figure capacity loss and capacity reduction breakdowns)   _____

5. Potential savings from downtime reduction (3 × 4)   _____

6. Percent of improvement possible through improved maintenance controls   _____

   0–10% if good maintenance practices are already in place

   10–20% if a basic manual system is already in place

   20–30% + if a weak or informal system is in place

7. **Total Downtime Cost Savings** (5 × 6)   _____

Labor Savings

Percentages:

1. Time wasted by personnel looking for
   spare equipment parts (use estimates
   below if actual is not known)   _____

no inventory system	=	15–25%
manual inventory system	=	10–20%
work order system and inventory system	=	5–15%
Computerized inventory and manual work order system	=	0–5%

2. Time spent looking for information
   about a work order   _____

manual work order system	=	5–15%
no work order system	=	10–20%

3. Time wasted by starting wrong
   priority work order   _____

manual work order system	=	0–5%
no work order system	=	5–10%

4. Time wasted by equipment not being
   ready to work on (still in production)   _____

      manual work
      order system   =   0–5%

      no work order
      system   =   10–15%

5. Total of all percentages of wasted time
   $(1 + 2 + 3 + 4)$ _____

6. Annual labor costs   _____

7. Total wasted labor dollars
   (5×6) _____

8. Select the percentage from the table
   below that best describes your
   maintenance organization: _____

      no work order
      or inventory
      system   =   75–100%

      manual work
      order system   =   50–75%

      manual work
      order and
      inventory system   =   30–50%

      Computerized
      inventory and
      manual work
      order system   =   25–40%

9. **Total Savings** (7×8)
   This will represent the projected
   savings from labor productivity.   _____

Inventory and Stores Savings

1. Estimated total maintenance inventory   _____

2. Estimated inventory reduction   _____

   no inventory
   system   =   15–30%

   manual system   =   5–15%

3. Estimated one-time inventory
   reduction (1 × 2)   _____

4. Estimated additional savings
   (3 × 30%)   _____

5. **Total Savings** (3 + 4)   _____

Major Outage and Overhaul Savings

1. Number of major outages and
   overhauls per year   _____

2. Average length (in days)   _____

3. Cost of equipment downtime in lost
   sales   _____

   [use daily downtime rate times
   total days of outages (total must
   be in downtime cost per day)]

4. Total estimated cost per year
   (1 × 2 × 3)   _____

5. Estimated savings percentage   _____

   no computerized
   work order
   system   =   5–10%

   pert system   =   3–8%

   pert system and
   inventory
   control system   =   2–5%

6. **Total Cost Savings** (4×5)   _____

Total Cost Savings

1. Equipment Downtime Costs   _____

2. Labor Savings   _____

3. Inventory Savings   _____

4. Major Outage Savings   _____

5. **Total Annual Savings**   _____

Additional Areas of Savings to Consider

1) *Warranty Costs*—These are costs that may be recovered from vendors for work that is done on equipment while it is still under warranty. These costs may or may not be recoverable if your plant personnel made the repairs.

2) *Quality Costs*—These would be quality costs that are directly caused, or could have been prevented by good maintenance practices. It is best to consult the quality control department to find what percentages of all quality problems are mainte-

nance related. This times the total scrap and rework costs for the plant for the year can be a sizable dollar value. The projected improvements and the related savings can then be calculated.

3) *Purchasing Costs*—These are premium costs paid for short delivery notices on emergency orders. These may include the expedited air freight, overnight delivery charges, or courier deliveries. Disciplined planning programs combined with preventive maintenance can dramatically reduce these costs.

4) *Overtime Reduction*—In many plants, maintenance overtime exceeds 20%. Plants with good maintenance controls will reduce this number to below 5%. This area can present some significant maintenance labor savings. However, a reduction in downtime for the equipment may also reduce the production overtime required to make up the lost production. This savings may be substantially higher than the maintenance costs.

5) *Energy Cost Reduction*—Properly maintained equipment requires less energy to operate. Some studies show this could be 5% of the total energy consumed by the plant.

Convincing the Corporation

This section of the program deals with winning and maintaining management support. It involves the following steps:

1. developing the plan

2. calculating the costs of the plan

3. calculating the savings from the plan

4. developing the cost/payback analysis

5. presenting the program.

By this time, if you have followed the outline of the book, you can accomplish steps 1–4. Step 2 would deal with the specific labor and supply (tools and materials) costs to put the various steps of the plan in place. More details on some of these costs will be found in the subsequent chapters dealing with the steps to the plan. The fifth step will be concentrated on in the rest of this chapter. It involves the maintenance manager becoming a "salesman" and being able to present the program in a bottom line-oriented manner.

It has been said about quality, and is true for maintenance, that "Good maintenance is not hard to do, it is just hard to sell." This is the area where most programs to improve maintenance fall short: the selling. In selling anything, the key is finding a need (increased profit, better quality, greater equipment effectiveness, etc.) on the part of the customer (management) and convincing him that your product (equipment maintenance management) meets that need. In order to accomplish this, you need to understand all you can about your product (maintenance) and the customer's (management's) need. Then it is your responsibility to use this knowledge to make the customer see the solution.

The maintenance manager is generally the individual that must start the sales process. There may be others in the company that start the process, but the maintenance manager must be involved if the project is to achieve long-term success. It must be remembered that T.P.M. is not a member of the program-of-the-month club; it takes long-term commitment.

Once the maintenance manager is committed to improvement, it is important to look at the long-term goals of the company and see where maintenance can complement and support these goals. The following steps can be used to help the maintenance manager accomplish this.

1) *Educate Your Management*—This is accomplished by documentation, such as letters, articles, case studies, and benchmarks from outside and inside sources. The communication should always be positive and related to costs benefit analysis, including increased profitability and facility benefits.

2) *Document Savings*—Any changes or improvements that are implemented should be monitored, and cost savings should be documented. The benchmarks that were established in the initial analysis can be used to confirm the savings and improvement.

3) *Think Results*—Always evaluate the results of any changes that are implemented. Measure these results against the goals that have been established. Even if the goal is not totally achieved, note the progress. Some benefit is better than none.

4) *Always Look for Ways to Make or Save Money*—This will always get management's attention. Any suggestions, even with slight improvements or savings, will be perceived as a positive indicator by management. This attitude always makes a difference when management is asked to invest in new maintenance projects.

5) *Write Articles and Speak to Groups*—This is positive reinforcement. It helps to stimulate others to action within the organization. If speaking is done outside the company, it can also attract new customers and clients. There is nothing more rewarding to management than to find that a major new account was influenced by meeting or listening to a knowledgeable member of the company.

6) *Keep Yourself and Others Current*—By staying current in the techniques and technology, there are several benefits. First, management respects professional and knowledgeable people. This allows the managers to be perceived as team members and contributors. Also, managers that have kept their knowledge

current find it easier to implement changes and can do so more quickly than if they need time to get themselves and the organization up to current techniques.

7) *Know Your Numbers*—It is important to always know your financial numbers. If you do not work from facts, you will never be successful. Working with the financial and budgeting managers can truly enhance the understanding of the numbers. This will allow the maintenance manager greater flexibility in preparing proposals for maintenance improvement projects.

8) *Be a Change Master*—It is important to stay close to the "cutting edge" in techniques and technologies. If you want to make progress, you need to make changes. Resistance to change is a difficult obstacle. Adapting the maintenance manager's attitude to being receptive to change makes it easier to change the organization. Any World Class organization must be able to make continuous and rapid improvement; and speed is an important attribute.

While many of the tips and techniques that have been described in this chapter are basic management concepts, they are the most overlooked items in maintenance organizations today. Only by concentrating on these basics will management support ever be gained and maintained. If the education and support of management is not given priority, the program will slow down within a year and die shortly after that time. It cannot be overstated: *gaining management support is the single most important step to beginning a successful Total Productive Maintenance program.*

5 Equipment Reliability

In this chapter we discuss the most common methods that maintenance departments use to ensure equipment reliability: preventive and predictive maintenance programs. These programs increase the availability of the equipment, but more importantly, they reduce the amount of reactive work that a maintenance department will have to perform. This allows the department to become more proactive and less reactive. This allows the maintenance organization the time to become more controlled, since they are no longer just firefighters.

Beyond the relaxation of pressure on the maintenance department, the results of the first formula in the equipment effectiveness computation begin to change. The decrease in breakdowns and downtime will increase the overall equipment effectiveness, producing the benefits mentioned earlier in the book. Some of the other benefits include:

support for Just-in-Time programs

support for Total Quality programs

creation of a successful environment for Employee Involvement programs

reduction of investment in capital assets

reduction of maintenance inventory levels

increase in corporate profits (the bottom line).

How effective are the preventive and predictive maintenance programs in the U.S.? As shown in Chapter 1, the majority of corporations are not satisfied with their preventive/predictive maintenance programs (P.M./P.D.M.). There are many reasons why the programs fail; however, the two most common reasons are the lack of management support and the failure to show results.

Reasons P.M./P.D.M. Programs Fail

The lack of management support is the problem that causes the end of the P.M./P.D.M. program. There are two sides to this problem. First, the program was never presented (sold) to management properly. Management may not have understood the resources that would be required or that the program would take time to produce results. This is usually the fault of the program champion, the sponsor, or manager. It is the individual who has the responsibility to present the program to upper management who must put all of the facts together: the initial costs for equipment and personnel, the problems getting time on the equipment for the service, and the time period that will pass before results will be shown. If management is sold a "pie in the sky" program, it will never succeed. P.M./P.D.M. programs take time and resources to set up and implement. Second, the results of the program have not been properly presented to management. Once the program has been implemented, upper management must be kept aware of the improvements in equipment reliability. Indicators such as downtime percentage, equipment effectiveness, and program costs versus program savings need to be tracked to ensure management support to the program. This is to ensure that as conditions in the plant improve, the true cause for the improvement is highlighted.

The second most common cause for P.M./P.D.M. program failures is the lack of actual results. This occurs when the program was designed. It was set up for the wrong equipment, or was not designed to prevent the true cause of failure. It is analogous to a doctor treating symptoms, but never the disease. The problem may be slowed down or go into remission, but it will eventually resurface. Then the comments such as "I thought the P.M. program was going to stop this!" will begin. It is only a short step to the loss of management support for the program. When management does not support the program, it disappears quickly. When the program is in its initial development, it is critical to have the true goal of ensuring equipment reliability in mind. The following material outlines some steps that should be taken during the initial setup of the program.

The first step is to perform an analysis of how to start the program. There are generally two choices: by department–location or by equipment. The department–location start means choosing the department or area of the plant that has the most problems with equipment downtime. This may be the bottleneck of the plant. Then the reasons for the breakdowns should be studied. Once the true causes of the breakdowns are found, the necessary corrective measures should be implemented. The second method of starting a P.M. program is by equipment. This would involve taking a list of the top ten equipment items that cause the most downtime in the plant (not in one location) and focusing the program on these items. The type of process, plant, or facility would determine which method should be used for a particular site.

In either case, results are shown on the equipment that is the primary concern to plant or facility management. The results, if properly documented, provide the necessary catalyst for management to become supportive of the program. Continued results will ensure continued support, if the results are properly documented.

Types of P.M. Techniques

The most common types of P.M./P.D.M. techniques are listed in Fig. 5–1. We will expand on each one, which will help in the selection of the proper technique to solve equipment problems.

Routines, Lubrication, Cleaning, Inspections, Etc.

This type of P.M. technique is the front line of defense against equipment problems. It is also one of the most critical in the development of a T.P.M. program. If the operators are to perform some maintenance tasks, they will be the basic tasks mentioned here. These tasks require few materials and few tools.

Routines may be small inspections and adjustments that are performed each day prior to equipment startup or right at shutdown. With a little training and coaching, these tasks can usually be performed by operations. Also included in the routines would be the task of tightening. This task prevents looseness of any parts, which in turn prevents any vibration and accelerated wear.

TYPES OF PREVENTIVE MAINTENANCE

1. Routine—Lubes, Cleaning, Inspections, etc.

2. Proactive Replacements & Scheduled Refurbishing

3. Predictive Maintenance

4. Condition-Based Maintenance

5. Reliability Engineering

Figure 5-1. The most common types of preventive maintenance.

The proper tools to perform the tightening should be supplied to the operator and kept at the equipment.

Lubrication is another task that can be performed by the operator. The best method is to color code the lubrication point and the lubricant, so there can be no mistake about what lubricant goes to what point. It is also necessary to specify how much lubricant goes in each point of application. In many cases, too much is as bad or worse than too little. This is a matter of training the operator to know what and how much goes where. In most cases, this is another duty that can be performed by operations.

Cleaning is one of the most beneficial tasks for a piece of equipment. While wiping off the equipment, the operator will spot many small problems before they can develop into larger problems. The operator can then write a work request for maintenance to correct the problem before it becomes a major one. Cleaning also keeps contamination from starting problems. Many equipment problems can be directly tied to contamination. Contaminated lubricant, and dirt or grit on machined surfaces, will cause accelerated wear and failures. By removing the contamination and repairing the source of the contamination, problems can be corrected in their infancy. The Japanese stress the cleaning of the equipment on a continual basis. They also stress that the initial cleaning, when starting a program, reveals many hidden conditions that will eventually result in equipment failures. The importance of initial and continuous cleaning cannot be overstated.

Closely related to cleaning is the inspection of the equipment. Inspection of the equipment can occur during any one of the previously mentioned activities. The operator can spot the problems and request maintenance to make the repair before any capacity reduction or loss occurs.

While this process looks good on paper, there are several problems related to operations performing this part of the maintenance program. The first is training. All operators will have to be trained to perform these tasks in accordance with the standard practices and procedures. This includes all related safety instructions and standards. Without the proper training, they will not perform the tasks properly, which will produce less than required results from the program. This could result in the failure of the program.

The hidden factor here is that very few companies in the U.S. provide adequate training for their maintenance personnel. The maintenance personnel are supposed to be the ones responsible for training the operators. If the maintenance personnel are not properly trained, what will be the skill and knowledge level of the operators? It will probably be insufficient to produce the results necessary to keep the program effective.

A second problem is the current status of the preventive maintenance program in most companies. If you take the time to look at one of your current P.M. checksheets, what do you find? Are they complete? Are they accurate? In most plants they are not! For example, if you have a P.M. checksheet for a chain (or belt) inspection, what does it say? Check the chain drive? That is what almost all plants have. However, it does not say *what* to check. It should list the items to check and give the specific parameters for the inspection. It should list things like checking the deflection in the slack side of the chain, and it should specify no more than 10% for the distance. It should list checking the sprocket alignment, the sprockets for hooked teeth, the chain for unusual wear on the inside or outside of the link plates. The list can and should go on and on. Most people like to stop about now and say: "My people are journeymen and I expect them to know this" or "My people are journeymen, they wouldn't want someone telling them what to do, they know how to do it." Both arguments are worthless. First of all, your journeymen may be doing the P.M. now, but they will not be doing it in the future. Operators, even with proper training, will need complete checklists. Second, the Japanese are big proponents of checklists. They have the philosophy that people can and do forget things. Checklists prevent this. Also, if checklists are properly detailed, they can become a training tool to help individuals find out-of-tolerance conditions. What is too hot? What is too long? What is too full? By knowing the tolerance, it can help both operations and maintenance personnel to become more effective troubleshooters and, more importantly, trouble preventers.

The previous tasks make up the majority of the T.P.M. tasks that operations personnel will perform. It must be remembered that the operations group never replaces the maintenance group; all they do is relieve some of the workload from maintenance. It is estimated that they will assume about 20% of the work that maintenance presently performs. Most of that work is in the activities listed above. Fig. 5–2 highlights the transfer of tasks. It should be noted that this does not result in an immediate reduction in the maintenance workforce. It merely frees up the mainte-

OPERATIONS GROUP RESPONSIBILITIES

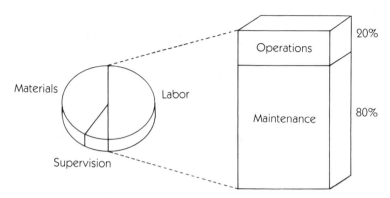

Figure 5-2. An operations group can assume about one-fifth of the work performed by the maintenance group.

nance personnel to work on tasks and projects that they cannot handle presently.

Proactive Replacements and Scheduled Refurbishing

Proactive replacements is the term used to cover shutdowns, outages, rebuilds, etc. These activities involve taking a piece of equipment or unit off line and overhauling it, replacing all worn or suspect components, and putting it back on line. It will then operate for a specified time period, with little or no maintenance-related downtime. This technique is seldom used without support from historical records or predictive techniques. The proactive replacement targets equipment components that are approaching the end of their life cycle. This can only be determined by using the techniques just mentioned. If these techniques are not used, then the cost effectiveness of the program will be seriously affected. If the parts' costs are too high due to replacing components that are not worn, then the program will lose support.

One note must be made: initial parts' costs will increase when the program is initiated. The first few overhauls on the equipment will attempt to restore the equipment to its original condition. Since most equipment is usually in poor condition when the program starts, an initial increase in the level of parts' replacement costs should be expected. As the equipment is restored through component replacement, then the parts' costs should become lower and almost fixed. If the proper presentation to management is made in the initial stages, the parts' costs will be accepted.

Predictive Maintenance

Predictive maintenance measures physical parameters against a known engineering limit in order to detect, analyze, and correct equipment problems before capacity reductions or losses occur. Predictive maintenance requires performing maintenance servicing when the equipment requires it. For example, consider a heat exchanger. If maintenance is performed when the equipment fails (becomes plugged), it takes longer to clean it out and restore it to service. If the cleaning is performed on a regular schedule—a fixed P.M. frequency—the cleaning takes less time and the equipment is out of service for a shorter period of time. If the exchanger is stopped up, the cleaning is much more difficult. However, the problem is: "Are we performing maintenance too soon or too late?"

This question is answered by applying predictive techniques. In the heat exchanger example, the predictive measurement could be the pressure drop through the exchanger. If the pressure drop exceeds a certain limit, then it is time to clean the exchanger. The exact limit on the pressure drop can be established by the vendor or the engineering department or even historical records. Once this has been established, then the cleaning occurs as needed. The frequency may be longer or shorter than the present P.M. frequency. However, the cleaning only occurs as required by the pressure drop.

Putting some estimates on the timing of the cleaning makes the benefits of predictive maintenance apparent. If the cleaning when the exchanger is clogged takes 16 hours, and the process is worth $5,000 per hour, and adding in the maintenance labor and parts, the cost for the cleaning is $80,000 per occurrence. If the system fails 6 times per year, the total cost would be (6×80) or $480,000.

If the fixed frequency P.M. program is used, a certain interval is specified. If the failures occur once every 2 months, then the cleaning may be set for every month. The cleaning takes only 2 hours, and the labor, material, and lost production cost is still at the $5,000 per hour level. Now the cost becomes $10,000 per cleaning or ($12 \times 10$) or $120,000 per year.

If predictive techniques such as the pressure drop across the exchanger is used, the cost is still $10,000 per cleaning, but the exchanger may only need to be cleaned 7 times per year. This translates into a cost of $70,000 per year. The predictive method has a tremendous cost advantage over the failure method. It still has a considerable advantage over the preventive method. In this case, the only equipment required is a pressure gauge on each side of the exchanger. This is a minimal expense compared to the savings.

The key to the predictive method is finding a physical parameter that will trend the failure of the equipment. The trend can then be used to predict the failure. Once the parameter has been found, it should have an upper and lower limit set, just like the upper and lower control limits in a Quality Control program. With the limit set, the current condition of the parameter measurement is monitored by taking visible readings on a route.

When the condition exceeds the upper control limit, then the required service is scheduled and performed. All of these actions take place before a failure occurs.

What are some of the major parameters that must be used to monitor equipment condition? The most common are:

1. vibration

2. shock–pulse

3. temperature

4. oil analysis

5. resistance.

While it is beyond the scope of this book to discuss these techniques in any detail, a short description of each follows.

Vibration Analysis—This technique monitors the frequency of vibration in rotating equipment. By choosing the proper frequency, all components in rotating equipment can be monitored and evaluated. All degradation of the component's condition can be trended and evaluated. Unusual wear can be detected and repaired before a failure occurs.

Shock–Pulse—This method detects the mechanical shock caused as equipment rotates. As the equipment deteriorates, the level of shock impulse increases. The level of shock can be monitored, and as it starts to reach a critical level, the equipment can be taken down and the defective component replaced before the failure occurs.

Temperature—This technique utilizes either thermographic cameras or infrared scanners to monitor heat. As the temperature of a unit changes, the image generated by the scanner changes. This alerts maintenance technicians to hot spots. Hot spots, particularly in electrical equipment, indicate a potential problem. Early detection allows the technicians to make adjustments or corrections before a failure occurs.

Oil Analysis—This technique utilizes analysis of both oil and wear particles to determine wear. This technique can be used on any system that contains oil, such as mechanical drives or fluid power units. The analysis can check the condition of the oil itself, ensuring that it has all its original properties. The analysis can also check the wear particles found in the fluid. The wear particles can indicate that bearings, gears, pump rotors, etc., are experiencing abnormal wear. This allows corrective action to be taken before a failure occurs.

Resistance—These checks allow the condition of key electrical components to be tested to ensure that they are not shorted out or otherwise damaged. This will permit the developing problems in motor coils and other resistive components to be isolated. This will again allow for proactive replacement of components before a complete failure occurs.

Condition-Based Maintenance

Condition-based maintenance is very similar to predictive maintenance, with the exception that workers are no longer sent out with hand-held devices to take measurements in the field. The sensors are mounted permanently on the equipment, and the signal is hard wired into a control room. Here all readings can be remotely checked, monitored, or perhaps integrated into a control system or a computerized maintenance management system.

Condition-based maintenance will still utilize some or all of the techniques that are mentioned in the predictive maintenance section. With the enhanced usage of the programmable logic controller (PLC's), real-time measurements from the systems in the field can be sent to the control room. Here the maintenance technician can monitor readings and actually perform some troubleshooting. For example, in a press and die operation, it is possible to monitor all hydraulic operations. If the system is properly designed, the maintenance technician can monitor and trend hydraulic pressures during the clamping and pressing operation. If he spots a falloff in pressure during the clamping, and there is no leak, he will realize that there is an internal leak in the system. By further checking each area in the system, he can isolate a cylinder or valve leak.

When all areas of a preventive maintenance program are examined, it is quickly seen that a good program utilizes all of the tools and techniques. Concentrating too extensively on one technique will produce a program that is not cost effective. The company that uses more diversity in the techniques will be able to design a cost-effective program.

Reliability Engineering

Reliability engineering is a subject that could take several volumes of material to properly cover, so this section will provide only a brief sketch. Reliability engineering is a discipline that is used to provide the final solution to problems that cannot be addressed by any of the above-mentioned techniques. It utilizes techniques such as redesign, retrofits, and even mathematical models to ensure equipment availability. When problems continue to occur with equipment, the engineer group can become involved. By utilizing mean time between failure and mean time to repair calculations at a component level, in most cases the problems can be found and possible solutions uncovered. If a component is prone to failure, then it can be isolated and studied. If materials need to be changed, or if the design needs to be changed, the engineering group can make the necessary changes and then continue to evaluate to ensure that the changes were effective.

Steps to Starting a Preventive Maintenance Program

Once the techniques have been analyzed and understood, it is necessary to begin putting the program together. How is this accomplished? The steps outlined in Fig. 5–3 should be followed.

1) *Determine Critical Units*—This is the step mentioned in Chapter 4 during the discussion of gaining management support for the program. It is choosing the equipment that will be included in the preventive maintenance program. While the goal is to eventually provide preventive maintenance service to all equipment, the critical or problem equipment should be the starting point. The critical units can be determined by

the highest amount of downtime

the highest lost production costs

the biggest quality problems.

STEPS FOR STARTING A PREVENTIVE MAINTENANCE PROGRAM

* Determine Critical Units

* Classify Units Into Types of Components

* Determine P.M. Procedures for each Type of Component

* Develop a Detailed Job Plan for each of the Procedures

* Determine a Schedule for each of the P.M. Tasks

Figure 5-3. The first steps to follow when implementing a preventive maintenance program.

The list could go on for a long time. The point is to make the selection based on the criteria of criticality specified by the operations or facilities that the maintenance organization services. This ensures that the correct equipment will be included in the program.

2) *Classify Equipment into Types of Components*—This step is designed to break the equipment down into its components. It is sometimes difficult to take a large piece of equipment and design a preventive maintenance program for the equipment in one step. By breaking the equipment down into components—such as hydraulic systems, electrical systems, mechanical drives, etc.—it is simpler to design a program to maintain it. The more defined the components are in this process, the easier it is to design the programs.

3) *Develop P.M. Procedures for Each Type of Component*— This step will develop specific types of services that are required for each component. These will include types of services that are common for the components. It would include all the techniques that apply to the components that were described in the previous section. There are general descriptions of the services that could serve as titles to the checklists and procedures that will be developed in the next step.

The information for the general lists of services that must be performed on each equipment item can be found in the following basic areas:

1. the manufacturer

2. equipment history

3. operators, craftsmen, supervisors.

All three of these methods should be used, and the results compared, to ensure full coverage for the equipment components. Each of the areas has its strengths and weaknesses. The manufacturers will tend to overmaintain the components. The equipment history relies on the past to predict the future. Any changes in equipment design, operating levels, or maintenance tools and techniques can alter the component's maintenance needs. Operators, craftworkers, and supervisors all tend to rely on their memories and can forget or overlook items. If all three methods are combined, the correct maintenance services can be listed.

4) *Develop a Job Plan for Each of the Procedures Listed*— This step takes each one of the individual lists of maintenance requirements and develops it into a specific step-by-step job plan. This is a critical and generally overlooked step in P.M. program development. The details on each P.M. inspection should include:

1. the craft required

2. the skill required for the craft (journeyman, apprentice, technician)

3. the tools required

4. equipment requirements

5. parts or spares required

6. detailed job instructions

7. estimated time to complete.

Most items on the above list require no explanation. Item 4 informs the craftworker of whether the equipment must be completely down before the inspection or service can be performed. If it is to be down during the P.M., the details should also specify how long it needs to be down. Item 6 is one of the most neglected steps in companies in the U.S. Most checklists or inspections are vague or incomplete. As mentioned in the section on types of P.M., without good details, items are overlooked or forgotten. If items are overlooked, failures or problems will result. This causes a lack of effectiveness of the program. A P.M. program that is not effective will lose management support and eventually fail completely. This spells disaster for the maintenance organization and ultimately for the company.

It has often been said that you only get one chance to implement a P.M. program per company (or per career), so paying attention to these details will help ensure that you do not fail.

5) *Determine the Schedule for Each of the Tasks*—The scheduling is a two-step process. It includes not only how long each task should take, but how often it should be performed. In determining how long it takes to perform a task, the estimate should consider:

time required to get tools and materials ready for the job

the travel time to get to the job

any safety, environmental, or hazardous materials restrictions

how long it actually takes to perform the task

how long it takes to clean up the area and put all tools and materials away.

If good estimating techniques are used, then the scheduling and completion of the tasks is much more accurate. However, there is one factor that can skew the schedules for P.M. programs: the time someone spends on the job performing work that is not on the P.M. inspection or service. As a craftsman services or inspects an equipment unit, he will occasionally find a problem that is beginning to develop. The question is how long should he spend correcting the problem before he asks for help, or writes a work order to correct the problem. This is generally a policy decision that should be made when starting a P.M. program.

There are two factors to consider when making the decision. The first is the time it would take to come back to the dispatching point and write a work order to have the work done. Depending on the geography of the plant, the travel time could be considerable; this should be taken into consideration. The other extreme is the damage a prolonged task would cause to the P.M. schedule. If a task is estimated for 4 hours, and it takes 8 hours to perform because of the other problems encountered, the schedule will suffer. Also, as the time is charged to the P.M. work order, performing what should be routine repairs and charging them to the P.M. charge number will inflate the P.M. costs and hide the true repair costs.

In most cases, companies will begin with a time limit of 1 hour. Any additional work could be performed up to the limit of 1 hour. If it is going to take longer than 1 hour, then the craftsman should come back and write a work order to perform the work. This will allow for planning and scheduling of the work which should make it more effective.

This leads into the second point—the scheduling of the P.M. program. Scheduling P.M. tasks depends on what type of P.M. is specified. For example, OSHA inspections, environmental equipment, and hazardous equipment all require certain servicing on specified and regulated intervals. If these P.M.'s are missed—for any reason—and the regulatory government agency checks, the company would be liable for fines. These would be classified as mandatory P.M.'s. They must be performed or else the company, equipment, or personnel will be harmed. These are usually fixed frequency P.M.'s and cannot be altered.

A second type is the pyramiding P.M. The pyramiding P.M. is

one that is due and is not completed in the allotted time window. Then, as a fixed frequency P.M., if it comes due a second time, another work order is issued. Thus, the work order pyramids. Nonpyramiding work orders are not issued a second time. They do not issue until the first work order is complete. Then, with fixed frequency P.M.'s, they slide the next due date based on the completion date, not the "true" next due date. Fig. 5–4 highlights the danger associated with this approach. With the pyramiding P.M. due on a fixed frequency, it is necessary to write a cancellation or missed notice on any uncompleted P.M.'s. In the example, 5 completions would be noted, with 2 being noted as missed. With the nonpyramiding P.M.'s, there are also five completions during the same time period. However, the nonpyramiding P.M.'s will show no missed tasks. When a failure occurs and the P.M. program is checked, the nonpyramiding P.M.'s will show no missed tasks. This leads the maintenance department to look somewhere else for the solution to the problem, when the real

PYRAMIDING

	Jan	Feb	Mar	Apr	May	June	July
Due	1st	1st	1st	1st	1st	1st	1st
Completed	15th	Missed	1st	7th	Missed	20th	15th

Due   7 times
Completed   5 times
Number missed   =   2

NONPYRAMIDING

	Jan	Feb	Mar	Apr	May	June	July
Due	1st	15th		5th	20th		1st
Completed	15th		5th	20th		1st	23rd

Due   5 times
Completed   5 times
Number missed   =   0

Figure 5-4. Nonpyramiding preventive maintenance programs can hide potential equipment problems.

fault is with the P.M. program. *Nonpyramiding P.M.'s can hide potential equipment problems.*

Another decision point in the program is whether to make the P.M.'s just inspections or make them task oriented. If they are just inspections, this means the inspector must come back to the dispatch point and write the work orders for someone else to go out and perform the work. This can lead to a rift between the inspectors and the rest of the maintenance workforce. The task-oriented P.M.'s instruct the inspectors to make minor repairs and allow time for them to do this. When making this decision, the future must be kept in mind. If complex tasks are specified, it makes turning the P.M.'s over to the operators more difficult. If the future direction is T.P.M., then the tasks must be set up and designed with that goal in mind. Otherwise, the transfer of any of the maintenance tasks to operations will necessitate a complete rewrite of the P.M. program.

P.M. Program Indicators

How can you monitor the effectiveness of the P.M. program? How can you know when the program needs adjustment to ensure effectiveness? There are several areas where observations can be made. They are

1. low equipment effectiveness (using the formula)

2. longer M.T.T.R. (mean time to repair)

3. maintenance-related quality problems

4. cost-per-repair increases

5. rapid decrease in the value of capital assets.

Low equipment effectiveness should be examined formula by formula. This could be particularly P.M. related when the availability is the lowering factor. If quality is the problem, then the quality part of the formula will highlight this point.

The longer M.T.T.R. indicates that when there is a failure or breakdown, it takes a longer time to repair. This indicates a more severe problem has been encountered. It takes longer to repair than a less serious problem that should have been found in its early stages by an effective P.M. program.

Comparing the quality problems to the equipment effectiveness formula can help to spot quality problems that are maintenance related. If the problems are related to routine services or P.M.'s, then the program's effectiveness is questionable. This indicator will allow corrective action to be taken.

The cost per repair is an indicator that shows repairs are more involved and are taking more parts and labor than they should, because of problems due to a failure or an advanced stage of deterioration.

The rapid deterioration of assets simply means that the equipment and facilities are not lasting as long as their design intends. The lack of maintenance contributes to higher than normal capital expenditures for equipment replacements. If the equipment is properly maintained, the effectiveness should be high enough to avoid purchasing replacements of backup equipment.

In development of any maintenance program, particularly T.P.M., it is essential to have a very effective P.M. program. This program eliminates emergency or firefighting maintenance, which is costly in labor and materials. It also disrupts the relationship between operations and maintenance. It is imperative that the program be as effective as possible. If the P.M. program is ineffective, the rest of the program will suffer and will eventually be discontinued. There have been many companies that have tried to replace their ineffective P.M. program with a T.P.M. program. If they cannot make a P.M. program effective, they will never make a T.P.M. program successful.

6 Maintenance Inventory Controls

If the other steps have been followed to this point, you now have a functioning work order system and preventive maintenance program. Without these two programs, a company will never have the discipline to utilize inventory controls. Good maintenance practices require a disciplined approach to inventory and purchasing. Without complete control of maintenance inventory, the next step of improving maintenance effectiveness can never be achieved. From this, all personnel must realize that just as operations and facilities require a good maintenance organization to support their efforts, maintenance requires a good inventory and purchasing organization to support their efforts. Poor inventory and purchasing practices are the single most common cause of poor maintenance productivity.

Consider that last statement for a moment. What are some of the most common delays that inventory and purchasing can create in a maintenance department?

1) *Craftworkers Waiting on Materials*—Think of the time that your employees spend waiting to get the materials to do their assigned job. This time can quickly add up to hours in just a single shift.

2) *Travel Time to Get Materials*—How much time is spent going to the job and then finding they need to go back to the stores to get parts? Or, how much time is spent getting the parts at the start of the shift? Are there long lines at the store window during the start of a shift?

3) *Time to Transport Materials*—Sometimes finding the materials is the first step; transporting the materials to the job can take a lot longer. This may involve finding a forklift or a truck to move the materials from stores or storage to the job site. This time can be even greater when there is a crew of workers assigned to the job. Many will wait, while the few make the move.

4) *Time Required to Identify Materials*—If stores materials lack numbering schemes for identification and location schemes for being able to find the materials, considerable time can be spent looking for the materials. Without numbering the parts with a clear identifier, it could easily be confusing to find the correct parts. One small difference can easily render a part unsuitable for the intended use; then all the travel and locating time begins again.

5) *Time Required to Find Substitute Materials*—It is difficult enough to find the right parts for a job, but when they are out of stock, it becomes important to find substitutes. If the parts are not quickly identified as substitutes, substantial time can be lost in finding these parts as well.

6) *Finding Parts in Alternative Store Rooms*—As organizations grow, it becomes necessary to maintain stores in remote locations to reduce the amount of travel time. This raises the problem of knowing what is carried or in stock in each of these locations. If the stock is out in one location, how much time does

it take to find out if it is in stock in another location? This is important to prevent reordering the item when an adequate supply may be on hand in a remote storeroom.

7) *Time to Prepare and Process a Purchase Order*—This can involve not just a considerable amount of time, but also a considerable cost. However, if a crew of employees is waiting on a part which has to be processed through purchasing, a considerable amount of time may be lost. This waste can be eliminated with proper controls.

8) *Time Lost Waiting on Other Crafts*—The inventory problems may be compounded if an organization works with strict craft lines. If one craft has the materials to start its part of the job, but one of the other crafts involved does not, delays occur for the entire job and all of the craftworkers involved. This can result in a tremendous amount of lost labor. If we compound these problems just mentioned with the others listed in Fig. 6–1, the

TYPICAL STORES COMPLAINTS FROM MAINTENANCE

1. I need this gasket but don't have the stock number.
2. I need a bearing like this. It is for the chiller pump on # 1 water system.
3. These are the same gaskets but they have different stock numbers for each of the stores.
4. If we would have planned this job, all of the parts could have been ordered at the same time, which would have saved money, and they would have all been ready when we wanted to do the job.
5. Using so many vendors, it is hard to track their deliveries and prices.
6. I know we have items we could liquidate, but I don't have the time to look for them.

Figure 6-1. Examples of inventory problems encountered by maintenance.

entire list of inventory problems becomes almost overwhelming. This is why inventory controls must be in place if maintenance effectiveness is to be achieved.

In order to be effective with the inventory system, it is necessary to understand how it should function and the information that should be maintained in the system. The following sections will address these points.

How a World Class Inventory Operates

In a World Class environment, the planner is the maintenance focus on the inventory department. The following work flow will provide insight into the planner's role.

1. When the work is identified, the planner determines what parts will be required.

2. Once the parts are determined, the planner issues the requisitions for the parts for the job for them to be kitted.

3. The storeroom personnel gather the materials together to a staging area, which may be a bin or a palette. The parts are clearly marked with the work order number for the job.

4. When the work order is issued to the craftsman, he goes to the staging area and picks up the parts. Alternatively, the storeroom personnel may deliver the parts to the job site the day or shift before the job is scheduled to begin.

5. The craftsmen use the parts. Any parts not in the kit may be requisitioned using the work order number. Any left-over parts are returned to the storeroom.

6. The storeroom personnel return the parts to stock and credit the correct work order with the returned part. They should never be allowed to credit a work order for a part that was not issued to it.

7. The items issued are compared to the items returned, and this comparison is used as a measure of planning performance.

8. If items are not in stock, the demand for the item triggers a reorder. The work order causing the demand will be put on hold till the parts are received. The purchase requisition is sent to purchasing. Purchasing will consolidate the purchase requests into purchase orders. This will allow purchasing to get the best price on any item from the vendor.

9. A second method for triggering a reorder is when the number of parts is below the specified minimum for that particular item. This reorder is less reactive than step number 8 and should be set to avoid any stock outages.

10. Another class of item that may require reordering is the nonstock item. This item is one that management has determined is not necessary to maintain in stock. The item may have a short lead time, it may be used infrequently with sufficient notice to order before use, or it may be a contract delivery with a local vendor. This item is still tracked through purchasing, but may or may not be issued through stores; this depends on the management philosophy regarding nonstock items.

11. Once an item is reordered, it must be tracked till received. The work can then be scheduled. If the item is

overdue for delivery, this should be noted and can become part of the vendor's performance rating. Once the item is received, the planner should be notified so the work order on hold can be scheduled as soon as convenient.

12. Periodically, the items in the store are counted to check for accuracy of the inventory. Any discrepancies are investigated and corrections are made. The quantities that are on hand should always be accurate to ensure high levels of service from the storerooms.

Information that Needs to be Maintained

The following are terms that apply to MRO (maintenance, repair, and overhaul) inventories. Some of the terms and definitions will vary slightly from production-based inventory systems.

The first term is "on-hand quantity." This should be the actual quantity that an individual would find if he walked to the storage bin or location and counted the number of items that are physically there.

The second term that should be noted is "quantity reserved or committed to work orders." This is not the number that is in the physical location. It is the number of items that are reserved for work orders which are presently scheduled or are ready to be scheduled. This number is critical because it lets the storeroom personnel know accurately how many items can be issued at the window.

The term that pulls the first two together is "available to issue." This term is the difference between the on-hand quantity and the reserved quantity. It is possible, due to incorrect stocking levels, to have a negative "available to issue." However, most stock systems only take the quantity to zero. This is why accurate planning is so important. If you hold extra materials for jobs they may not be required for, it prevents you from doing work that may have a higher priority.

Quantity on order is the number of items that are presently on order, but not yet received. The other information that goes with this number is the due date. This allows the planner to note when the parts required are due to be received. If the parts are not received by their promised date, then the purchasing department should get a notice to follow up with the vendor to find out why the items have not been delivered.

Two other fields commonly tracked in inventory are maximum on-hand quantity and minimum on-hand quantity. The minimum on-hand quantity is the lowest number of items that you want in stock before an order is received. This quantity is different from reorder level (or reorder point). The reorder point is when the order should be placed. In theory, it should be placed before the minimum quantity is reached. If the minimum quantity is reached before an order is placed, it is likely that a stockout will occur. The reorder point should be set so that the lead time will allow the quantity to reach the minimum level just as the order is being received. What should trigger the reorder? It is when the quantity available to issue reaches the reorder point. This is important, since the reorder cannot be driven off of the quantity on hand. As was explained previously, the reserves are deducted from the quantity on hand to get the true number available to issue. If the reorder was based on the on-hand quantity, then continual delays in scheduled maintenance work would result from the excessive stockouts. The maximum quantity should be the minimum plus one order quantity.

The order quantity is the number of items ordered when a reorder is placed. This number plus the minimum will provide the maximum. The maximum should never be exceeded. However, if poor planning disciplines or poor integration between a maintenance system and a stores system exist, it is possible for this rule to be violated. When this occurs, then the company is keeping excess stores items. When this occurs, the holding costs, storage costs, taxes, space requirements, etc., are all higher than necessary. This is classified as inventory waste.

Another item to track in the inventory is the last issue date. This information helps to monitor slow-moving and obsolete items. While it is true that some large maintenance spares may never move, it is good to review the inventory to ensure that items with no movement within a specified time period are reviewed. This time period will vary from one company to the next, but usually falls between 6–12 months. If the inventory is kept in separate storerooms, such as in area or multiple storerooms, the stores personnel need the ability to query other stores locations to see if they have the part in stock before any orders are placed. If they do, then some companies perform material transfers. This allows any demands for materials made while there is stockout in one location to be met. This reduces or eliminates any downtime that would otherwise be incurred. In order to ensure that the quantities the stores are keeping are accurate, there is a process called a physical count. This is where the stores personnel actually physically count the number of all stock items. This is usually performed by location to keep the cost of travel low and to enhance productivity. The actual count is reconciled to what the card system or computer says is the count. Any adjustments that are made, either positive or negative, are charged or credited in accordance with the company's policy.

There is a variation of the physical count process called a

cycle count. This is where stores items are classified, usually by amount of activity or by cost, and are assigned a code. All "A" items may be counted once a year. All "B" items may be counted once every six months. All "C" items might be counted once every three months. Once the frequency is established, then the items are divided so that an equal amount can be counted every week. This allows for a smooth workload for the stores personnel and still keeps accurate track of the inventory.

Some common performance information that should be kept for maintenance by stores are number of stockouts, number of issues, number of returns, and number of adjustments, just to name a few. Comparing the number of issues to the number of returns allows for a measure of planning effectiveness. Comparing the number of planned issues to the number of unplanned issues will also help to isolate problems with planning. The number of stockouts could be used to help find items that have incorrect max–min levels and to reorder quantities specified. Monitoring the number of adjustments can help spot pilferage or poor storeroom disciplines.

Purchasing information is a step beyond the storeroom data. The purchasing department must not only have all the part information, but must also keep vendor information. This would include the vendor contact's name, address, phone number, etc. Purchasing will also have to track the vendor's performance. This includes things like number of late orders, and number of orders shipped incomplete or incorrect. Purchasing may track overages, underages, poor quality, damage, etc. All of this will be factored into a performance rating for the supplier or vendor. The purchasing function has the responsibility to ensure that the prices for the items being purchased are the best. This requires that maintenance avoid unplanned work since this can dramatically affect the pricing of an item. Emergency purchases prevent being able to obtain volume discounts, price breaks on shipping, and planned or interval purchases.

If maintenance demands are planned and forecasted, then purchasing can consolidate all requests for items from one vendor into a multiple line item purchase order. This will also reduce the cost of purchasing items. Real costs that are sometimes ignored by maintenance personnel are purchasing costs. Consider for a moment what it costs to process a purchase order within your company. Is it $50? $100? $200? Depending on the company, single line item purchase orders can cost anywhere in this range to process. Multiple line item purchase orders can cost between ½–¾ this amount. So, the logical thing to do, in order to reduce costs, is to improve the forecasting and order policies for maintenance. Yet, in many companies, it is not uncommon to find purchase orders for $10, $20, or $30. This is a waste that must be eliminated.

If these costs are controlled, there are some other areas to consider, as follows:

1. standardization of plant equipment

2. standardized supplies

3. proper storeroom locations

4. reduction and elimination of obsolete parts

5. elimination of spoilage.

The standardization of the plant equipment provides maintenance with far more than added inventory controls, including reduced training, reduced drawings, etc. However, inventory is the primary concern in this chapter, so it is this focus that will be explored. The reduction of inventory levels occurs because all of the equipment has basically the same parts. So instead of carrying 15 sets of spares for 15 presses, you may only carry 5 sets. Since maintenance repair and overhaul frequencies should be staggered to prevent simultaneous downtime, you should never need more than a few sets of spares at any given time. This can amount to tremendous savings for any size company.

The standardization of supplies also prevents overstocking of similar supplies. We will consider one example. Lubricants are one of the most overstocked supplies carried at any site. Consider for a moment: how many different types and grades of lubricants do you carry from different manufacturers? The answer in most cases exceeds 20. However, how many companies have gone to their suppliers with their lubrication needs and asked the supplier how many would be required if they were given a sole-source contract for all the lubricants in the plant? In most cases, it would be below 10 and, in some cases, 5 or less. By reducing the number of lubricants to that level, think of how many interchange charts this eliminates. Consider the number of lubricant-related failures due to mixing incompatible lubricants it will eliminate. These may be failures that are presently going unsolved. The lubricants, due to the chemical additives, will coagulate or lose their viscosity when mixed. The problems can go virtually undetected unless oil analysis is performed. This problem can be compounded when poor training has been provided and the craftworkers think that all lubricants are the same. The problem can be further compounded if lubrication duties are turned over to some of the operations personnel, since they are less likely than the maintenance craftworkers to have the required training. This example with the lubricants is just a sample of the many problems that could be reduced or eliminated through part standardization.

Proper storeroom locations are critical to the productivity of the maintenance workforce and to quick service to prevent unplanned downtime. If standard supplies and critical spares are not stored in a location that is geographically close to the equipment they will be used on, there is time lost due to travel and finding and transporting spares. In some cases, where storage close to the equipment is impossible, deliveries have to be made. However, deliveries only work when good planning and scheduling disciplines are practiced. The problem is further compounded in a true T.P.M. environment, since all parts that the operators use are to be stored at the equipment. This would inflate the stores levels, without a delivery system. So it is best to utilize area stores where necessary and supplement this with a delivery system where it is not possible to have area stores. Some companies have eliminated up to 10% of their total manpower expenditures by utilizing area stores and delivery systems.

The elimination of obsolete parts is also a key element to effective maintenance stores controls. There is a penalty for every spare part carried in stock. It includes the storage cost, such as space, heat, lighting, labor to move the part in and out of stock, etc. These costs can grow to be considerable. If 25% of the inventory is obsolete, then a considerable charge is being made to the company for parts no longer required. Obsolete parts are sometimes kept in the stores when equipment has been retrofit, sold, scrapped, etc. If accurate spares records have been kept, the spares should also be removed or scrapped at the same time as the equipment. This avoids the collection of junk in the storeroom.

Spoilage is generally caused by buying too much of a spare or supply. In normal practice, this is an error on the part of purchasing. They may see the opportunity to get a volume discount on an item and so order a larger volume. Then the item sits on the shelf and spoils. It is important to use historical trends and patterns when purchasing maintenance spares. Volume usage of spares can vary dramatically from month to month. For example, the stocking level of an item may be 24; however, maintenance has not used any in the last 11 months. The problem goes without investigation, so it may not be discovered that maintenance only uses the item on a certain overhaul or retrofit. Further study may show that the retrofit only occurs once per year, and at that time it takes all 24 of the items. Depending on when the inventory evaluation occurred, the decision may be to raise or lower the stocking level—and both solutions would be a mistake. It is best, if good planning is enforced, to try to have a Just-in-Time maintenance inventory. When this occurs, spoilage will be eliminated. In order to assist the manager in controlling the maintenance inventory, it is necessary to break the inventory into categories. The most accepted method is the ABC classification (see Fig. 6–2). In the ABC classification, "A" items are items that make up

ABC ANALYSIS

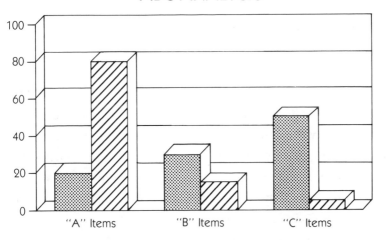

▒ % of Total Number ▨ % of Total Value

Figure 6-2. The "ABC" classification of inventory.

approximately 20% of the total inventory, but are worth 80% of the total inventory valuation. "B" items are another 30% of the inventory value, but are worth about 15% of the total inventory valuation. The "C" items make up the remaining 50% of the total items inventory, but only comprise 5% of the total inventory value. It becomes the priority to closely control the "A" and "B" items, since they make up the majority of the maintenance costs. "C" items can often be covered by blanket orders with vendors, allowing less labor and control expense to be invested by the company.

Maintenance Requirements

What does maintenance require from an inventory system? Some of the basic needs are as follows:

real-time parts information

equipment "where used" listing

projected delivery dates

part usage by cost center.

While most of these are self-explanatory, and are in addition to some of the items that were discussed earlier, there are several that are very important. The first is the access to real-time parts information. Maintenance personnel need to know what parts are on hand at the exact time they are making the request. There are some manual and computerized systems that are only up-dated periodically. The update may occur once per day or, in some cases, once per week. This means that in between the times that the information is updated, the information is inaccurate. So if a key equipment item is down and requires a part, someone could be sent to an area stores to get the part, while a crew is disassembling the equipment. Once it is discovered that the part is not there, either the equipment stays down while the part is being ordered, or it is placed back together until the part is received. In either case, there is a waste of maintenance labor and probably unnecessary equipment downtime. Maintenance needs to know what parts and part quantities are in the storeroom at the exact time of the inquiry. The second area requiring attention is the "where used" listing. This is a list of all equipment the particular part or spare is used for. This facilitates the removal of all related stores items when a piece of equipment is sold, scrapped, or otherwise removed from the plant; it also allows for the planners to find equipment that uses the same parts or components. This enables planners to check the service life of identical components installed on different equipment items, so that failure patterns are checked and it is possible to determine if it is a component problem or an equipment problem. This feature also allows for emergency borrowing of parts from equipment that is shut down or not in use. Planners need projected delivery dates in order to project the possible start dates of work orders. In many cases, special order parts are required. Without these dates, scheduling work orders becomes "guess work." This will damage the credibility of the maintenance organization. With good projected delivery dates, maintenance planning and scheduling becomes more accurate and reliable. The parts usage by cost center is more for a financial audit, but can provide some useful trending information. If one department or cost area is using too many parts, the maintenance practices and policies for this area can be investigated. If the parts are not being used, then what is the labor for the area being used for? This analysis provides some interesting audit of maintenance policies. As with all other parts of the maintenance, the stores/purchasing function needs a method of measuring performance. Some of the most common indicators include

1. turnover

2. stockouts

3. service level

4. inventory accuracy

5. slow-moving parts

6. backorders

7. purchasing service levels.

The turnover indicator shows how many times the number of items you have in stock turn over, i.e., used in a given time period. To be accurate, each one of the categories (ABC) should be evaluated separately. The number of turns will vary greatly depending on whether the items are "A" (high dollar, low volume), "B" (mid-range value, medium volume), or "C" (low value, high volume). When companies develop fixed rules, such as "all items must turn three times per year," they are making a major mistake. This type of rule impacts the "A" and "B" items, which are the items that will cause the highest amount of equipment downtime when they are not available. The average number of turns for maintenance spares for the entire inventory should be 1 per year. "C" items will be higher, while "A" items may not turn at all during a particular year. The percentage of stockouts indicates the true level of inventory service. The stockouts occur when there is a demand for a part that is currently out of stock. This indicates a problem with the max–min levels, the reorder points, or another inventory setpoint. The stockout can result in lost maintenance productivity or lost production. Both of these costs need to be considered when setting the order and stocking levels for parts. The stockouts should be less than 5%. The service level, which is just the inverse of the stockouts, is the percentage of time that the parts are carried in the stores. The service level should be greater than 95%. Inventory accuracy is the comparison of what is actually on hand to what the records (computer or manual) say is on hand. This report can only be generated after a physical count of the items has been taken. The more accurate the inventory count, the better control the company has over its inventory. If the accuracy is low, it shows that there are issues and returns being made to stock that are not being recorded. When this occurs, the stockouts and service levels will be affected. In order to be truly effective, inventory accuracy should be greater than 95%. Slow-moving parts are stores items that are not turned within a specified period of time. The time depends on the company—ranging from 6 months to 1 year. This standard should be closely applied to all "C" items and some "B" items. It is assumed, by definition, that most "A" items would be slow moving. However, all slow-moving parts should be evaluated at least annually to see if adjustments can be made that will allow for optimum stocking levels to be achieved. Backorders and vendor delivery performance are indicators of how the suppliers or vendors are doing in servicing the company's needs. The vendor should have a high on-time delivery rating, with very few orders either late or on backorder. Again, if the planners have scheduled work tentatively, based on promised delivery dates, then the productivity of the labor force could be impacted. Vendors with low delivery performance and high backorder numbers should be replaced with more responsive vendors. Purchasing performance is measured by service levels; this includes measures such as timely order cycles, updated delivery information, rapid processing of transactions, etc. Purchasing's responsiveness to maintenance needs is required to support the effort from stores. Without timely ordering, inventory levels will have to be maintained at an artificially high level to prevent stockouts and low service levels. All parts of the company impact one another in this cycle. If the true customer for all of these departments (operations/production/facilities) is not satisfied, then the profit picture becomes very bleak. In conclusion of this chapter, it must be finally stated: just as maintenance exists to service operations, stores and purchasing exist to service maintenance. If this order is observed and understood, the company has taken a large step to ensure its profitability and ultimately its competitiveness.

7 Improving Maintenance Effectiveness

The techniques and controls described in this chapter build on the controls and disciplines mentioned in the previous chapters. The ability to improve maintenance effectiveness is built on enforcing the basic disciplines mentioned previously. Without a commitment to implementing the previous programs, the techniques described in this chapter will not be effectively utilized by any organization. Assuming the controls and disciplines are in place, you can begin improving maintenance effectiveness by examining your craftworkers.

Examine the crafts at your particular site. Do the craftworkers take pride in their work? Is the quality level of their work high? High levels of pride and craftsmanship are common in organizations where management supports the craftworkers. However, when examining your particular plant, do you find that management provides the craftworkers with the right tools for the job?

the proper materials for the job?

materials that are properly located to eliminate lost travel time?

equipment that is scheduled down so that they can work on it without arguing with operations?

clear job instructions detailing what is to be done and how it is to be done?

clear explanations as to the objectives and goals of the work assignment?

If these essentials are provided to the maintenance technicians, then they can take more pride in their assignments, since they feel like part of the team. Management has provided the necessary information for the technicians to make logical decisions while performing the job. In this way, decision making is pushed to the lowest level. The job may be planned, but some decisions will still have to be made while the job is in progress. Good, clear information allows the technicians to do this.

Typically, in most companies, inadequate planning techniques and the lack of coordinated information produce the following delays and wasted productivity for the craft technicians:

waiting for or questioning instructions

looking for supervisors to make a decision

checking out the job since the instructions were incomplete or missing making multiple trips to the storeroom since the parts were not planned, or the wrong parts were planned

trying to find the right tools for the right job

waiting for approval to finish the job since it is going to take longer than originally thought

having too many or too few craftworkers, causing one group to look for more resources and the other to hide the excess resources they have.

In any of the above cases, the problem is lack of coordination and utilization of the resources that are available. It is a waste of the available hands-on time, i.e., the time the technicians actually have their hands on the job in progress. This is the time you are really paying for. So labor productivity is one of the most important maintenance resources. How do other departments within a company control their assigned resources? For example, how does the

financial department manage resources?

production department manage resources?

engineering department manage resources?

Each one of these departments utilizes planning and scheduling techniques to maximize its effectiveness. Now ask yourself: "If all these other areas utilize planning and scheduling techniques to optimize their resources, why don't we use the same techniques to control maintenance resources?"

An important point that is overlooked in maintenance management is maintenance planning—the step that brings all the basic techniques discussed to this point together to optimize the utilization of all resources. Without maintenance planning, the organization will never be cost effective.

Maintenance planning requires a dedicated individual(s) to plan and schedule all maintenance activities. A good ratio of maintenance technicians to planners ranges from 15:1 to 20:1. If the numbers are higher than that, the effectiveness of the planning program is in jeopardy.

What is involved in planning maintenance activities? It is basically an effort to avoid all of the losses mentioned earlier. All work of a nonemergency nature, including P.M. and P.D.M. inspections and services, should have a detailed job plan. The job plan includes

all required materials

the craft and skill specified

the required number of craftsmen

a listing of all required tools

a list of any nonstandard equipment

a detailed description of all job steps

descriptions of all safety requirements

any OSHA, EPA, and other federal or state requirements.

As mentioned previously, it is the *maintenance planner's* responsibility to ensure that all of this information is provided for each nonemergency work order. This is also the reason why the planner/craftworker ratio is so important. If a planner is to provide all of this information, he will be limited to the number of work orders that he can process in a given time period. If the ratio is not kept reasonable, then the equality of the plan suffers and the effectiveness of the planning program becomes questionable. This has led to the termination of the majority of failed planning programs in U.S. industry.

The first-line supervisor in maintenance is then responsible for seeing that the job is executed according to the job plan provided by the planner. This involves the supervisor being on the floor with his crew at least 6 out of the 8 hours available on the shift. The supervisor should not be tied to administrative or paperwork functions for more than 2 out of the 8 hours available on the shift. If he is, then he becomes nothing more than a high-paid clerk. It is important to allow him proper control of his crews, since he is responsible for their activities.

Scheduling brings the entire activity together. A maintenance schedule is most effective when performed on a *weekly* basis. Some companies try to schedule on a daily, monthly, or some other timeframe. However, they either lack the control or are too confining to be effective. Examining the weekly schedule, it is the most effective, since it allows flexibility, yet still has enough control to avoid wasting resources. For example, a daily schedule can be disrupted when made out 16 hours before the starting time of the schedule. Once the schedule is set, and the planners go home for the day, any breakdowns or emergencies that occur before they come back will disrupt the schedule. On a day-to-day basis, the schedule will be unreliable and inaccurate.

The weekly schedule is far more accurate because it allows for emergencies and other schedule interruptions. In a previous chapter, we discussed the capacity work scenario. The amount of emergency work, small interruptions, P.M.'s, and other events can be tracked and averaged on a weekly basis. Consider the following example:

10 men × 40 hours		=	400 hours
2 contract employees × 40 hours		=	80 hours
5 O.T. shifts × 8 hours		=	40 hours
Total available to schedule		=	520 hours

Deductions:

30% emergency work (0.3 × 520)	=	156 hours
5% absenteeism	=	26 hours
P.M. work (20%)	=	104 hours
Total deductions	=	286 hours

$$
\begin{aligned}
\text{Total to schedule} \quad &= \text{available} \quad - \text{deductions} \\
&= 520 \text{ hours} \quad - 286 \text{ hours} \\
&= 234 \text{ hours.}
\end{aligned}
$$

If the planner would schedule 234 hours of work from the backlog for the week, the crews would have a high probability of completing the tasks.

The same technique was described in the section on maintenance backlogs and staffing levels. If these techniques are used to set the amount of work to schedule, scheduling accuracy of greater than 95% can be achieved.

Scheduling Flows

How should the maintenance scheduling process take place? This depends on the organizational structure; but the steps below should be followed.

1. The planner gathers any incompleted work that is outstanding at the end of the week.

2. The planner calculates the craft capacity for the next week.

3. The planner will deduct outstanding (incomplete) work from the craft capacity.

4. The amount of craft capacity that is left is the total number of hours that can be taken from the craft backlog.

5. Based on priority, date needed, the equipment availability, the planner selects the jobs from the backlog for scheduling. It must be noted that any work that is to be put on the schedule for the next week must be ready to schedule. This means that all parts, tools, outside contractors, rental equipment, etc., must be ready. Any work that is put on the schedule before it is ready will do nothing more than create lost productivity and wasted resources.

6. The planner will continue to select work from the backlog, matching the labor resource requirements to the availability. Once the availability is matched, the schedule is complete. The planner may wish to put several additional jobs in a category of optional work. This is in the event that the number of emergencies are lower than expected or there is a change in production schedules, restricting the equipment that will be available to work on during the week.

7. The tentative schedule is presented to the maintenance manager for review and maintenance approval.

8. The maintenance manager should meet with the production/operations/facility manager by Thursday of the week before the schedule is to start. The production/operations/facility manager may want to rearrange some priorities, add some jobs, cancel some jobs, etc. This

interchange between the managers should finalize the schedule.

9. The finalized schedule is presented to the planner. The planner begins the printing of all work orders, picks parts lists (for the stores), makes contractor notifications, and contacts equipment rental agencies.

10. The planner then puts the information about each work order into a packet and delivers it along with the next week's schedule to the supervisor that is responsible for the work, by noon on Friday.

11. The supervisor has time to look over the work schedule and resolve any questions before the end of the day on Friday. This allows the supervisor to prepare the order in which the work is to be executed during the following week.

12. It must be noted that the schedule does not tell the supervisor the order in which the work is to be done. This is the supervisor's responsibility. It is also the supervisor's responsibility to match the craftworkers to the job. The supervisor is familiar with the skill level of each assigned employee. It is the supervisor's task to match the employee's skill level with the particular job.

13. As the week progresses, the supervisor will turn in all work orders that have been completed to the planner. The planner will complete the record keeping (this may be a clerical job, depending on the resources available).

14. The planner monitors the progress of the schedule completion, and by Thursday is ready to begin the schedule for the next week. The process begins to repeat.

There are some variations and adjustments that must be made to make this scenario work for all companies. Multiple planners will require that the schedule is coordinated. Multiple crafts require the same additional coordination. Multiple operations managers require multiple meetings and maybe multiple schedules. Area maintenance will work slightly differently than centralized maintenance. However, if the basic principles discussed here are applied, and the basic disciplines are enforced by all parties involved, effective scheduling will occur.

By observing the requirements for good schedules, it can quickly be seen that no maintenance schedule will ever be effective if good planning is not enforced. Effective schedules require that each work order have

accurate craft requirements

accurate materials requirements

accurate contractor requirements

accurate equipment/tools requirements

accurate date needed, priority, etc.

Again the point must be highlighted: without effective planning programs, effective maintenance scheduling is just a dream.

In review, some of the following points should be highlighted.

All maintenance work should be scheduled.

All maintenance work should be planned by experienced technicians.

The schedule should use capacity planning techniques.

No work should be scheduled until it is ready.

The work should be processed as backlog, weekly, and then daily work.

The planner has responsibility for the backlog and weekly schedule.

The supervisor has responsibility for the daily schedule.

8 Maintenance Automation

This chapter examines the computerized maintenance management system and the ability to interface maintenance information to other organizational systems. The chapter begins with the benefits achieved by computerizing, and concludes with the selection and implementation phases.

Why Computerize?

Fig. 8–1 details the four main reasons for computerizing maintenance management. We will explore each one in more detail. First, Computerized Maintenance Management Systems (CMMS) provide a structure for enforcing maintenance disciplines. Every CMMS, whether purchased from a vendor or developed in-house, has a certain philosophy of how maintenance should operate. It has certain information that is collected on a work order, certain stores and purchasing procedures, certain

GOALS FOR COMPUTERIZATION

* Provide Vehicle for Enforcing Maintenance Disciplines

* Provide Faster, More Accurate Record Keeping Capabilities

* Provide "Snapshot" Analysis of Maintenance Information

* Provide a Method of Integrating Maintenance with Other Information Systems

Figure 8-1. Four prime reasons for computerizing maintenance management.

reports, etc. The structure and philosophy of the CMMS will become the structure and philosophy of how your company is going to manage maintenance. The importance of selecting the right CMMS becomes very clear at this point.

It is at this point that the "chicken and egg" question comes up for the maintenance manager: "Do I need a good manual system or paperwork system before I computerize?" Many consulting firms will answer "yes," and then spend months helping a company to develop all of the paperwork to manage maintenance. When your CMMS is implemented, you find that you have to change many of the paperwork policies that the manual system put into place in order to use the computerized system. If you are going to use a CMMS, implement it, enforce the disciplines that it requires, and save (what can be) considerable time and money. If the selection of the right system is made initially, it is a waste of time to develop a manual system first.

The CMMS should also provide a faster, more accurate record-keeping methodology. Paperwork systems generally have only one point for data entry. A CMMS may have many points of data entry. This allows for faster record keeping. Since the systems are usually definitive on the data they are requiring, the data going into the system are checked for validity. This ensures that the data are more accurate and complete than what is found in a paperwork system. The accuracy and timeliness of the data help provide the managers with better information on which to base their decisions.

The CMMS also should have the ability to provide snapshot or summary reports that allow a manager to see the data in a concise, meaningful form. This eliminates the browsing through pages of data trying to find a trend or even a particular fact. This is an area where you may have to watch some off-the-shelf CMMS packages, since some do not provide analysis or exception-type reports. Some only provide lists, which impair the usefulness of the system.

A computerized system, if properly selected, also has the ability to interface to other computerized systems in the company. This may include an existing stores or purchasing package, a payroll package, a general ledger package, production scheduling package, etc. The interface can eliminate the passing of paper or redundant data entry points. The savings in clerical support alone can be substantial. The data accuracy and reliability can prove to be beneficial in improving maintenance communications with other departments that may already be computerized.

The problem with CMMS is that, in many organizations, the maintenance manager is not involved in the selection process. In some cases, he does not know anything about the software until someone brings the computer and software and puts them on his desk. In this case, the maintenance manager and the maintenance organization will not be prepared to use the CMMS. The mindset is that just by giving the manager a computer and software, it will make him more effective. This is not the case. It is similar to walking into the maintenance organization and giving it a collection of musical instruments. This will not immediately make it a symphony orchestra. It takes time, practice, and dedication. It takes the same to implement and use maintenance software.

What makes a good CMMS? A good CMMS must support all maintenance activities. This means that it must have

work orders

preventive maintenance

inventory

purchasing

personnel

management reporting.

As has been discussed in the book to this point, all of the above are required to make a maint   nce organization function properly. So any CMMS shoul'   above modules. However, just having these m   ugh. It is how these modules work that is t   on. For example, some systems can   w keystrokes; other systems   of screens and data to do a goc   it is at this level of testing and exam.   should try a CMMS before the final purc   wever, each organization should keep its needs in   ne software is evaluated.

If a packaged CMMS is c   sidered, what type of vendor can a company expect to do business with? The vendors fall into four basic groups:

1. software developers

2. nonmaintenance consultants

3. maintenance consultants

4. maintenance integrators.

The software developers are programmers that have developed a software package for maintenance. They usually have little, if any, maintenance experience. The software reflects this, since it is usually very well written, with good coding practices. The software will lack a definite maintenance philosophy and, in many cases, necessary maintenance disciplines. The vendor is very willing to take its "base" package and customize it to meet the company's needs. This can be very expensive, and should be considered as a last resort.

Nonmaintenance consultants are firms that may have production, accounting, inventory, purchasing, or other backgrounds that have decided to branch into the maintenance area. What are some of the signs of this? The package will reflect their area of expertise. If it is production control, the package will lack the flexibility that is required for maintenance planning and scheduling. If the background is financial, then the system concentrates on the financial aspects of maintenance. It will lack the necessary functionality to be effective in managing maintenance.

The maintenance consultants that have a package generally will enforce the necessary maintenance disciplines. However, the coding for the software is generally poor. The software will not take full advantage of the programming language or the hardware platform. Since these groups evolved from maintenance consultants, they tend to try making projects out of each sale. They may try to put staff people on site for extended periods of time. It is not true for all companies, but it is an area to watch.

The maintenance integrators are the full-service vendors in the marketplace. They have the blend of maintenance and software expertise that allows for the premium products in the marketplace. They can provide assistance during the implementation process in the area of systems and maintenance technologies. These are the best vendors to deal with in the marketplace. The only disadvantage in dealing with these groups is that they are relatively more expensive than the other vendors. However, sometimes you get what you pay for.

Selection Process

What is the best method to select maintenance software? The process should begin with a company defining its needs. Begin by asking what you want the system to do for maintenance. Then expand the question into other areas and parts of the organization that are going to be affected by the CMMS. Once these needs are listed, develop them into a requirements document. The requirements document may be something as simple as a checklist. It can also expand into a request for information, a request for proposal, or a request for quotation. However, keep in mind that the more extensive your requirements are, the more the system is going to cost.

A major mistake that many companies make is to go out and look for maintenance software without knowing what they want the software to do for them. When this occurs, the wrong software is purchased. The company will then expend unnecessary resources trying to make the software match the organization or trying to make the organization match the software. It becomes a long and ultimately frustrating endeavor that never realizes its potential. Always know what your requirements are before looking for software.

This becomes especially important if the end goal of your whole program is a T.P.M. implementation. If the system that is purchased does not support some type of work request and approval process, it becomes difficult for operations personnel to enter and track their work orders. Also, if the CMMS does not support multiple users, then the operations people will never get to use it to request their work.

What are the most important selection factors for a CMMS? Fig. 8–2 shows the results of one survey. As can be quickly seen, the relationship between the vendor of the CMMS and the purchasing company takes the first three spots on the survey. The vendor, particularly the maintenance support, is critical to gaining acceptance of the system within the maintenance organization of the purchasing company. This even carries further down the list into user training. How the vendor conducts the training, and how effective it is, will make a major difference in how quickly the system is proven effective.

What are the most common types of problems encountered during the selection and implementation of the CMMS? Fig. 8–3 highlights the most common problems. For example, included in the wrong type of vendor is the vendor's software. It is possible to choose a vendor that does not fit your organization. When this occurs, either it never meets the real needs of the organization, or it requires too many support people to make it function. In either case, the CMMS does not function the way it was expected to, and the client company is dissatisfied. It is a matter of putting up with the shortcomings of the software or changing it and purchasing another one.

Management support is essential, but this hurdle should have been cleared earlier in the program. However, good communication will help upper management to understand what is going on during the implementation. The lack of willingness to make organizational changes can be a serious problem. In some cases, the organization needs to change to adapt to the maintenance system. No one has the perfect organization, and since a compa-

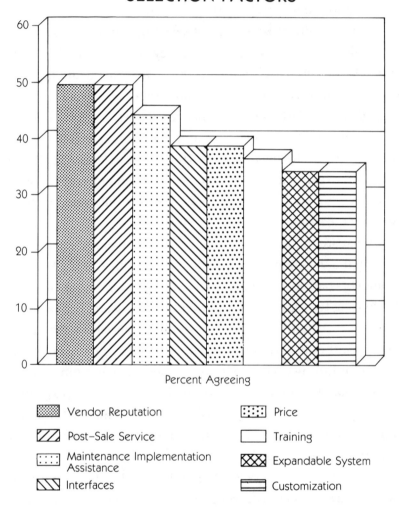

Figure 8-2. The most important considerations in selecting a computerized maintenance management system (CMMS).

SELECTION AND IMPLEMENTATION PITFALLS

* Wrong Type of Vendor Doesn't Meet Needs

* No Management Support

* Wrong Software/Hardware

* No After-Sales Support

* Lack of Willingness to Make Organizational Changes

* Choosing In-House Development

Figure 8-3. Problems that are commonly encountered during the selection and implementation of a CMMS.

ny is trying to improve by purchasing a CMMS, it should not be resistant to making changes that will help it to be more effective in utilizing the tool it has purchased.

The last point in Fig. 8–3 is the thought of in-house development. While there are many companies who have opted for in-house development, it is a poor choice. The reason is that you end up with one of two extremes. Either the system never is as fully functional as it could be, which means it does not do as much as it could for the maintenance organization; or it is so complex and complicated that no one in the organization wants to use it. For companies considering in-house development, they should know that it takes twice as long to design and implement a system as it does to purchase one. However, even worse, the development and internal support costs will be over ten times more than for a purchased package.

If a company cannot find anything on the CMMS market that meets its needs, it should negotiate with a vendor to purchase the vendor's source code and customize that. At least it will be starting with 50–60% of the system already in place.

The bottom line on in-house development is that it is usually undertaken by a company or individual within a company with an *ego problem*—one who feels that he can do it better than any of the 300 or more vendors that have had or still have CMMS packages on the market. This has caused more problems and cost more money to companies than anyone would dare to admit.

System Implementation

What is the best method for implementing CMMS software? It is generally a five-step process, as shown in Fig. 8–4. This is a matter of preparing all of the data for entry into the system. Equipment and inventory numbering schemes were discussed earlier in the book, but now they must conform to the limits of the software. If you have a 20-digit equipment number and the software only accepts a 10-digit number, you have a problem. This is one of the areas that must be carefully considered during the selection process. In order to ensure that the software matches the data, you should ask the vendor to give you the fields and their sizes and to make up forms to gather the data. If the forms look just like the screens that the data are going to be entered into, you can save time and money in the later stages of implementation.

The second step—the installation of the hardware and software—is a critical step in the success of the program. Many small companies think that just installing the hardware and software is as easy as plugging in the computer. In an industrial setting, power is unclean, subject to fluctuation and frequent outages, etc. All of these problems will create data corruption, software failures, and general user headaches with the entire project. This can all be avoided by using good protection devices on all computer equipment. Backup power supplies can be utilized in areas where frequent power failures occur. Industrial-grade computers can be utilized in any areas where environmental conditions are poor. While they are somewhat more expensive, they can avoid loss of a system or the data, which would be much more expensive.

Frequent backup of data is an important protection against loss of data. In some cases, a power spike, or a crash of the computer, destroys the data in the computer. Without backups, all the information is lost. With backups, it is a matter of reloading the software and data, and the system is ready to run again. How often should the systems be backed up? At least once per day, with intervals of twice per day during the initial data-loading process. Backups, however, remain one of the most neglected areas of any CMMS in operation, particularly in the PC or PC-LAN environments.

If the CMMS is going to operate on a minicomputer or a mainframe computer, the computer support people for the company will probably take care of installation and backups. The maintenance manager and those involved in the use of the system should still know how (in theory) the backups and system maintenance are scheduled and performed. This will eliminate the "black box" mystique that these systems tend to develop.

Most of the systems come with effective security levels that allow for multiple users on a system. This keeps data and some system functions restricted to those who need access to them. However, security is only as good as the users want it to be. If managers leave their passwords open where others can get access to them, then the security of the CMMS is compromised. By following the recommendations of the vendor, adequate security should be assured.

The third step is the data entry phase of the implementation. If the data have been gathered in step one, on the forms that look

STEPS TO SUCCESSFUL IMPLEMENTATIONS

1. UPDATE CURRENT RECORDS

A. Equipment Nameplate

Numbering schemes

B. Preventive Maintenance

Relate to equipment/components

May have to write tasks

C. Stores and Purchasing

Numbering schemes

Usage history

On-hand quantities and locations

D. Personnel Histories

Current skill levels, etc.

E. Work Order Files

Accurate history

2. SYSTEM INSTALLATION

A. Hardware

PC protection

PC/LAN system protection

Backups

User security

Terminal locations

Printers

Mini/Mainframe

B. Software

PC/LAN

Mini/Mainframe

Understand the process

Figure 8-4. The five-step process for implementing CMMS software.

3. DATA ENTRY

   A. Electronic Transfer

      Cost from vendor

   B. Manual Information (Step 1)

      Forms

      Temporary clerical support

   C. Consistency

      Upper/lower case

4. SYSTEM INTRODUCTIONS

   A. Ongoing

      Pay envelopes

      Bulletin boards

      Crew meetings

      Supervisor contacts

      Company newsletter

   B. Characters

      (Nonthreatening)

5. PERSONNEL TRAINING

   A. Right Staff People

      Don't train plant manager

   B. Right Type of Training

      Job/task oriented

      Hands-on, real world

   C. Right Conditions

      Your site

      Vendor's site

      Neutral ground

Figure 8-4. The five-step process for implementing CMMS software.
(continued)

like the screens, then you can hire temporary data entry clerks to put the data into the system. This eliminates having the maintenance staff loaded down with this job. In fact, most maintenance staffs are already working hard just to stay up with their present workload, so that making them do the data entry takes a long time. If the data entry takes too long, then the enthusiasm for the CMMS project starts to wane. If data entry professionals are used, then the process is quick and relatively without an increase in the maintenance staff loading.

If the present data are stored in a computer, whether in another CMMS or a home-made database, it is possible for the vendor of the new CMMS to electronically transfer your data from the old to the new system. This can save a tremendous amount of resources during the data entry phase. However, this will depend on the vendor of the CMMS and how you have kept your data in the present system.

One last point on data entry: do not use any data that are suspected of being inaccurate or unreliable. The "garbage in—garbage out" maxim is very applicable when it comes to computer data. Make sure that all data that are put into the system are accurate and reliable, otherwise none of the output from the CMMS will be.

While system introduction is listed as the fourth step, it actually is ongoing throughout the selection and implementation process. Depending on the organization, starting a CMMS program can be quite a cultural shock. Good communication can help to limit the amount of resistance there will be to change. One vendor has been successful by inventing cartoon characters to help the employees identify with the system. It is hard to feel threatened by a cartoon character.

Some of the other methods listed in step four can be useful for ongoing communication. Each site will have to determine the method it will use to keep the lines of communication open.

The training of the users is the final step for the implementation. This is one of the most overlooked areas of system implementation. It is the area that companies will omit when submitting project budgets. This is a major mistake since the training will determine how quickly the system will be implemented and how fully it will be utilized. Yet, despite many warnings, companies will still say that they can do it better, or even do without training. This is absurd. Think about some of the most popular courses offered at adult education classes. They all involve the use of database and spreadsheet packages that cost under $1,000. Yet these courses are almost always full. The software comes with user manuals, but still people want to learn more. The people want to become *proficient* at using the software. The only way to do this is to take training. Now if individuals want to spend money learning how to effectively use their personal software, why can't companies spend money teaching their employees how to use software costing thousands of times more than personal software? It does not make much sense, yet companies continue to fall into this trap.

The training should be the right type as well. It should be real-world problems, using the company's data. The trainee should have to use the software, not just listen to a lecture. It should be in a classroom/workshop-type setting, free from interruptions and distractions of the employee's normal job. Good training programs should be available to all who are going to be users of the software, if the implementation is going to be successful.

Additional Areas of Concern

Some additional areas of concern include underestimating the time to implement, and overselling the benefits of computerizing to management. The time it takes to gather the data and enter them into the CMMS can be considerable. Most consulting

WORLD CLASS VERSUS PRESENT STATUS COST BENEFITS

WHAT TYPE OF SAVINGS CAN BE REALIZED?

1. Labor

     50% increase in productivity

     15–30% savings in labor costs

2. Maintenance Materials

     15–25% reduction in inventory

3. Tools

     20% reduction in repair and carrying costs

4. Supplies and Miscellaneous

     15% reduction in maintenance expenditures

(All figures are industry averages)

Figure 8-5. Substantial savings can be realized with a CMMS.

firms will estimate 1 hour per record to gather the data and put them in the CMMS. Now, of course, the present condition of the data can have an effect on the time, but this rule of thumb is fairly accurate. So, if a company has 10,000 inventory items, and 2,000 pieces of equipment, then the time becomes rather considerable. If it only dedicates 1 person to enter the data, it will take years to get the job done. The correct approach is to see the number of records, calculate the time needed to get the data entered in, and then dedicate the resources to get the job done. This keeps upper management satisfied, when dollars and timeframes are consistent and understood from the beginning.

The second problem is overselling management on the benefits of computerizing maintenance. There are industrial averages that can help in this (such as Fig. 8–5), but each organization is different. The size of the savings and the improvements must be calculated with full understanding of the present organizational conditions and constraints. As was seen earlier in the book, just computerizing will not solve some problems. Each manager must understand the present condition of the organization and its future course before putting together a cost justification for a CMMS.

CMMS is a great tool for maintenance, but it is just that—a tool. It is not a magic cure for all the problems that ail a maintenance organization. If a manager keeps the goals outlined in this chapter in focus during the selection and implementation process, then he should have a successful project.

9 Training for C.A.T.'s

Training for C.A.T.'s? What is a C.A.T.? It was once said that to improve maintenance, the first step would be to change the name of maintenance, otherwise it would never get any respect. Each company would have to make its own acronym for maintenance, but, as a sample, I use Capacity Assurance Technician, or C.A.T. for short. While the idea may seem insignificant, it has made a difference in companies that have tried it.

Beyond just a title change, maintenance workforces are in need of education. This has been evident even more recently in view of educational statistics that show 13% of all U.S. 17-year-olds are illiterate, and only 1% are illiterate in Japan and Germany. Also, the graduation rate in the U.S. is only 73%, while it is 94% or more in Japan and Germany. This adds more credibility to the claim that the American workforce needs more training. How effective will these people be when they enter the workforce?

144

What added burden of training will be put on American companies?

In the matter of industrial training, consider how much training companies are funding at the present time. When was the last time that your craftworkers were exposed to meaningful training? Was it 6 months ago? 1 year? 2 years? Longer? Most training groups will admit that if you have not updated your craft technicians' skills in the last 18 months, their skills are outdated. But how are training programs organized? What subjects should be included? The following sections will answer these questions.

Types of Training Programs

The various options to fill the need for a highly trained workforce are listed in Fig. 9–1. The easiest really involves no formal training, but just hiring the personnel that are already trained. These people are usually available from other companies

TRAINING ALTERNATIVES

* Hire Trained Personnel

* Vocational Schools

* Train In-House

* Train at Vocational Schools

* Vendor Training

* Colleges and Universities

* Continuing Education Programs

Figure 9-1. There are several ways to assure that the maintenance workforce possesses the skills required to perform efficiently.

in your geographical area. What are some of the discussion points for this option? First, while these people may be skilled, their knowledge will be general, not specific to your equipment, process, organization, etc. So even if the need for formal training is eliminated, there still is the need for some on-the-job training to help them adapt to their new surroundings and job assignments. A second disadvantage to this approach is the expense. A good journeyman in any craft is making the maximum wages for the geographic area. So to hire someone of any quality away from his present position, it will take a better opportunity (usually financial). So the issue becomes: is it more cost effective to hire the expertise at a premium cost, or to hire someone at an apprentice level and train him to become a journeyman? The correct answer will vary from area to area, but the issue should be evaluated closely before any final decision is made.

The second method is to hire a trained apprentice from a vocational training program. There are presently vocational programs that produce good apprentice-level individuals that can quickly be productive. What are the advantages of this method? First, they come to the job with a theoretical knowledge, with only some hands-on experience in a lab-type of a setting. This is not a real handicap, since they have not had the opportunity to learn any bad habits. These people will be able to start at entry-level compensation and still possess some of the basic skills to make a contribution to the work effort. This helps keep the initial employment costs low.

What are the disadvantages of hiring people with vocational training? First, their knowledge is not complete or may not provide competency in some areas. The level of proficiency that they bring to the job will depend greatly on the background of the instructor that they had in the program at school. If the instructor's approach was just theory, then all they bring to the job is textbook knowledge. If the instructor had real-world shop experience, then the student will have a higher level of job-related skills. However, this approach is still not a cure-all since it will still require the use of an on-the-job training program or some form of in-house training program. If the in-house training program is not used to complete the training of the vocational employees, then they may take years to "learn by osmosis" the things they need to know to be effective.

The third option is in-house training. This is where the company sets up and maintains its own training center for maintenance. When all of the options have been considered, experience shows that this is the best method for training craftworkers. The negative is that it is also the most expensive. While it may be the most expensive, it is the most controlled, targeted to meet specific needs, and provides the highest quality employees in the least amount of time. However, this raises the following questions.

Who does the training?

What facilities are required?

What materials should be used?

Are the employees compensated for their time?

What is the true cost?

First, who is going to do the training? This question will receive many different responses depending on who you ask in the organization. Rather than explore all of the options, the best answer is: a craftsman or supervisor with good job experience and the ability to teach. The instructors should have the respect of the trainees for their job knowledge, and should teach the material in a manner that shows practical application. This allows the trainees to make specific application of the material, thus avoiding the usual question: "Why do I need this information to do my job?" The enthusiasm usually runs high when the students know that the material will further enrich their knowledge base and ultimately their own self-worth.

The facilities that are required will include classrooms and lab setups. The classrooms should be used for the lecture type of program. They should have available overhead projectors, slide projectors, video machines, and even, in some cases, some PC computers for any computer-based training. The basics that should be considered are adequate lighting, heat, and air conditioning. The classrooms should be set up with tables for writing and using reference materials. The lab setups should use equipment that can be used to highlight the material covered in the classroom. Being able to teach the theory and then go to the lab and show it in real-world action makes it meaningful to the trainees and provides the necessary reinforcement to make the retention rate high.

The materials that can be used include correspondence or canned types of training materials, textbooks, video tape programs, in-house developed material, or materials that are custom developed by outside companies. Each of the materials has an application in the training program. The best results are usually obtained by using a blend of the above materials. The vendors for these materials are usually listed in any of the magazines that service the maintenance market, or they may be found at some of the trade shows that have a maintenance focus. Each of these training approaches has its advantages and disadvantages, but this would lead into a discussion of training philosophy and methodologies, which would be beyond the scope of this text.

During in-house training, should an employee be compensated for attending training? The answer again varies from company to company, with the collective bargaining agreement being (in most cases) the deciding factor. If employees are in pay-for-knowledge types of programs, they have an incentive to want the training. It may be easier to implement training programs where the attendance is unpaid. However, the smart companies will realize that they benefit from the training also, so they will work with the employees to find an equitable solution. Where companies have a pay-for-time-on-the-job type of agreement, it is harder to get employees to want either paid or unpaid training. The entire company philosophy on training, pay-for-skill, and career advancement potential must be examined when investigating training compensation.

Without exception, it can be said that a quality, in-house training program is more expensive than any other form of training. However, if it is properly conducted, it is the most effective. It requires resource commitment from upper management, with the realization that there will be no short-term payback. The classrooms, labs, equipment, and training materials will be a considerable expense. However, the results from in-house programs are faster and of better quality than any of the other methods.

There are specific training programs that can be set up and conducted at local vocational schools. These courses will use the votech school's facilities, but will use the material you specify. The course can be set up several different ways. One may be just for your company. Another may be set up for just a particular topic, but the school can open the course up to any companies that want to attend. The difference is the cost. A course specifically for your company means that you will pay the entire cost of the course, minus whatever state or federal funds that might be available. A course open to the general public will allow the cost to be spread over all the participants.

A sticking point to the vocational classes can be the instructor. In some cases, the instructor may lack the necessary real-world experience to be effective with your craftworkers. Instructors with only theoretical backgrounds that cannot make the transition to application of the material will find themselves challenged frequently by the craftworkers. This may prevent the transfer of knowledge that is required to judge a training program effective. Again, training the craftworkers requires an instructor with real job experience.

Vendor training involves either sending craftworkers to a location where a vendor has a school, or having the vendor come to your plant and conduct the course. Some points of interest about vendor training include the question of sales-type presentations. In some cases, the vendors will send out technical people, and they can do an excellent job conducting the training program. However, in some cases, they will send out a sales representative, who tends to slant the presentation toward a description of the vendor's products. This can turn a good program into a total waste of time and money. It is always prudent to investigate the background of the presenter before the training program is set up.

The vendor type of training can provide some good equipment- or product-specific training. In most cases, the vendor's technical personnel are competent and very qualified. They should be able to answer most questions and use common examples to reinforce their presentation. The training materials are of good quality, since in many cases they double as the sales materials also. While this type of training usually has a narrow focus, it should never be overlooked for specific equipment or problem-solving training.

Colleges and universities can be used for some training programs, however, most of this training will be theoretical. This is why colleges and universities are good for advanced electronics, fluid power, or mechanics training; and the point should be made that the term "advanced" was used. Many companies use this method to train their instrument technicians or electronic repairmen. It is effective and can provide the background necessary for maintaining the advanced technological equipment in the modern plant.

Continuing education programs are specific seminar types of programs that may address a variety of different topics. These programs are usually 2 or 3 days in length and cost hundreds of dollars for someone to attend. However, if you need this particular program, it may be the only place to get the information, without developing your own course. One additional point for consideration when you need many people trained is to have the seminar leader conduct the seminar in house for your company. This usually involves the seminar leader's fees and expenses, but is far more economical than sending several of your employees to a public seminar. With some advanced effort, the seminar leaders can sometimes tailor their course to your specific needs.

Course Outlines

What types of topics should be covered in a good maintenance training program? Fig. 9–2 covers some of the basic topics. The mechanics training should not include bridge stress calculations, but should involve subjects such as

bearing care and maintenance

V-belt care and maintenance

chain care and maintenance

gear care and maintenance

screw threads and drives

mechanical fasteners.

TRAINING TOPICS

* Mechanics

    "How To"

* Fluid Power

    Hydraulics

    Pneumatics

* Electricity

    AC/DC

    Low Voltage

* Electronics

    Board versus Component Repair

Figure 9-2. Examples of topics that should be taught in a maintenance training program.

The list can become quite long. The point is to teach them what they need to know to do their jobs correctly. Some companies will spend the time to do a job-needs analysis. This includes studying a particular job and seeing exactly what tasks are involved for each job. Each job is then broken down into tasks. Each task is then studied to see what knowledge is required to perform it. This required knowledge is then written and developed into a training program.

Consider the fluid power course mentioned in Fig. 9–2. This course could be taught from a design standpoint. This is theoretical but not practical for the maintenance technicians. However, consider some of these subjects:

fluid reservoirs—care and maintenance

flow control valves—care and maintenance

directional control valves—care and maintenance

hydraulic pumps—care and maintenance

troubleshooting fluid power components.

The topics mentioned here can also continue to be expanded. It is important to make the material job related, teaching just enough theory to help the technicians understand the hows and whys of repair and troubleshooting.

Electricity requires the most theory of any of the industrial maintenance-related topics. However, even *it* can be simplified by teaching component functionality and operations. Circuit operation and troubleshooting techniques must also be included, but this helps the technicians see the application of the theory.

In electronics, the structure of the training program depends on what you want the technicians to be able to do. If you want them to be able to troubleshoot and replace components on a board, it requires extensive training. On the other hand, if you are going to require them to just be able to change the defective board and send it out for repair, it requires somewhat less training.

It all relates to knowing your needs. That is why a needs assessment or needs analysis is a prerequisite to developing a cost-effective training program. Otherwise, you will spend too much or too little on training. Either mistake will be a waste of resources.

Operator Training

With the ultimate goal of T.P.M. being to transfer some of the maintenance tasks over to operations, the training program can be useful for this also. If the training materials are developed properly, the maintenance technicians can then use the materials to help the operators understand the whys and hows of their tasks. If the materials are not correctly developed, then the effort will have to be duplicated to develop the operators' training materials. More on this topic will be presented in Chapter 11.

With technology changing so rapidly, training of the maintenance and operations personnel will become critical. The companies that invest in training will be competitive. Those that choose not to train will eventually be out of business.

10 Total Asset Management

If the plan has been followed up to this point, then the maintenance organization is beginning to develop a large number of data. The question becomes: "How can these data be used to optimize the effectiveness of the maintenance function in the company?" By this time, the traditional barriers between maintenance and operations should be disappearing. The real source of this traditional, long-standing problem is the lack of communication. Each group has difficulty in justifying its side of the discussion in terms that the other can relate to. In fact, operations managers may still have some difficulty in understanding how much priority should be given to maintenance activities in relation to their own goals and objectives. This chapter discusses techniques that can be used to translate all factors into a common language that all in the company can understand.

The search for one language that is understood throughout the company can always be reduced to "financial justification."

To effectively utilize this language, it will be necessary to understand how maintenance contributes to the financial standing of the company. One statement must be clear to all involved: *nearly all maintenance decisions are cost benefit decisions.* This means that all decisions that involve maintenance are considered by asking: "What is the company going to get in return for spending this money on maintenance?" The answer may be reduced depreciation of assets, increased production, or higher quality. The problem lies in expanding our point of reference so that we can see the whole picture.

The cost of maintenance can be broken down into the following major categories:

labor costs

material costs

overhead costs

function loss or function reduction costs.

The labor costs are the costs for having the maintenance technicians on staff. This will include all related expenses including overtime, benefits, etc. Each job will have the cost of the number of employees times their hourly rate plus any incentives. This cost makes up approximately 50% of all direct maintenance expenditures.

The material costs include all parts and equipment replacement expenses controlled by maintenance. The costs of all parts or equipment components used on a job should be tracked through the work order system. Material costs make up approximately 40% of all direct maintenance expenditures.

Overhead costs include the clerical and staff support for maintenance. Typical overhead costs range from about 15 to 25% of the usual maintenance labor costs.

The function loss or function reduction costs are the hidden costs, or costs which are difficult to ascertain. It is estimated that these costs will range from 2:1 to as high as 15:1 per maintenance activity. So when a maintenance task costs $10,000 in labor and materials, function loss or function reduction costs could cause the true total costs of the incident to run as high as $30,000–$160,000. This makes a tremendous difference in looking at the cost of a maintenance action.

In trying to calculate this type of cost, what are the problems that are encountered? The problems revolve around the accuracy of the data that we are working with. How do we judge how much to spend and when to spend it? The second problem is being objective when gathering these data. Without definite data collection disciplines, the data may be distorted, favoring either the operations or maintenance viewpoint. It is important to accurately collect the data to ensure that the company gets what it pays for when it comes to maintaining its assets.

The savings or benefits from maintenance revolve around cost avoidance. This means that spending the money on maintenance avoids incurring an additional and greater equipment repair cost. What are some of the areas where costs can be calculated? They would include the following:

preventive/predictive maintenance which avoids function loss breakdowns

repair/replacement maintenance which corrects function loss breakdowns

improving/restoring function reduction breakdowns

compliance with governmental regulations (environmental, safety, hazardous materials, etc.)

cosmetic appearance for operators, customers, etc.

The preventive/predictive maintenance program requires the expenditure of resources with the calculated probability of avoiding function loss breakdowns. This expenditure must be weighed against the probable cost of the breakdown. This keeps the preventive/predictive program operating in a cost-effective manner.

The repair/replacement function is usually in response to a problem that was noted during the inspection and monitoring step of the preventive/predictive maintenance program. Once a potential problem is noted, the repair or replacement can be planned to cause minimal interruptions to the operations department, and still prevent or correct a breakdown.

The restoration of a function loss breakdown will usually involve some cleaning or replacement of a component on a trend of decreasing efficiency. This could be a task such as cleaning pumps with decreasing efficiency or restoring the full operational capacity to production equipment. This is an overlooked area of maintenance improvement.

The compliance with governmental regulations can also be a function of maintenance. For example, one municipality was fined over 1 million dollars for an environmental violation. The cause of the violation was ultimately traced to a pump that was improperly maintained. Fines are a small part of the total picture when damage to the environment, employees' health, or other damage due to improper handling of hazardous materials are included.

In the T.P.M. mindset, the environment that the employee must work in, the impression that the operation gives to the customer, and the public's opinion of the company may not easily translate into dollars, but it can be a factor. For example, it is estimated that over 20% of all manufacturing costs are related to quality problems. Understanding how maintenance affects quality costs is important to be able to quantify any costs in this area.

It is necessary to put the costs-versus-benefit discussion in a form that all parties involved can understand. Fig. 10–1 shows the material discussed above in a graphic format. The figure shows that the decision for maintenance would be made based not on what is best for the operations group, nor on what is best for the maintenance group, but on what is the lowest combined costs. This is the effective "bottom line" for the company. This is the type of decision that companies must make if they are to optimize their resources.

How does a company go about collecting the information required to perform some of these analyses? It starts with assigning a cost to downtime. It may be useful to use the financial or

COST EFFECTIVE MAINTENANCE

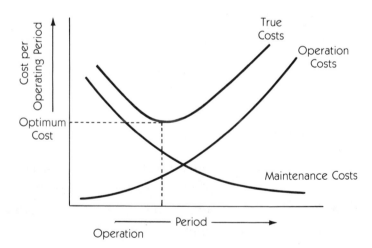

Figure 10-1. Maintenance timing decisions can be made by establishing the lowest cost ratio between maintenance and operations.

accounting departments to find out what 1 hour or 1 shift of lost production is worth for 1 piece of equipment. This might include lost sales, employee salaries and overhead, the cost to make up lost production (if it can be made up), and any measurable depreciation to the assets. The figures coming from the financial department will usually be conservative, but will not be disputed by other parts of the organization.

With these figures agreed to, it is necessary to understand the maintenance costs involved. These may include the labor, material or supply, and miscellaneous costs that will be incurred due to the repair or the failure. Both costs may be needed to compare an overhaul to a run, to failure approach, to maintenance. Additional costs that may be incurred should also be calculated. These may include the hazardous materials, EPA, OSHA, or safety considerations.

Once the total cost is understood, some interesting problems can be considered. The following three problems address scenarios in three areas. There are many other problems that each company may identify and so use this technique to justify and solve. The three main areas are

preventive maintenance frequencies

critical spares analysis

normal maintenance parts and supplies.

For an example of examining P.M. frequencies, a piece of equipment common to most plants or facilities will be selected: a centrifugal pump. This pump may be pumping a product or moving cooling water. The point is it will have a value to its service. Setting a price on the value gives a reference from which to start. If the value is $100 per hour, then this allows the managers to solve the following problem. Suppose it costs

$1,500 for parts and labor to overhaul the pump. There is no downtime cost since there is a standby pump available. The pump performance is measured, and it is found that after 4000 hours of operation, it loses 5% of its capacity. It is assumed that the drop is linear and continues to be so throughout the life of the pump. The question is asked: "when is it cost effective to remove the pump from service and clean the rotor?"

The problem is solved by calculating the amount of maintenance cost versus lost performance cost per hour. The two are combined to give the lowest total cost. The techniques to perform this analysis are illustrated in Figs. 10-2–10-5. In Fig. 10-2, the maintenance cost is calculated. The mistake of only considering maintenance costs is highlighted in this particular example. If only the maintenance costs were considered, it would be advisable to delay servicing the pump for as long as possible.

The point in Fig. 10-3 is that if you delay the service, the amount of lost production cost is increasing linearly. However, the difference between Figs. 10-3 and 10-4 is that the latter takes into consideration that the performance falloff is triangular and not the total area volume of the rectangle.

The summary of the problem is in Fig. 10-5, which plots the falling maintenance cost against the increasing cost of the lost performance. These two factors added together will give the total of the true costs. The decision can then be made on lowest true cost, which in this problem would indicate that the maintenance action should be performed every 1500 hours of actual run time.

The problem can become more complex since this problem has no penalty cost for the downtime required for the maintenance action. If it would cost additional monies for the downtime, then this column would have to be added. However, in the example just given, the only downtime that would likely be

BASIC MAINTENANCE COST CALCULATION

Repair Cost = $1,500

If the pump was serviced once every 100 hours, the cost would be:

$$\frac{1500}{100} = \$15.00/hr$$

If the pump was serviced once every 500 hours, the cost would be:

$$\frac{1500}{500} = \$3.00/hr$$

The table version would be:

Service Frequency (Hours)	Maintenance Cost (Dollars)
100	15.00
500	3.00
1000	1.50
1500	1.00
2000	0.75
2500	0.60
3000	0.50
3500	0.43
4000	0.38

Figure 10-2. A method for calculating the maintenance cost of a pump.

incurred is when the repair was made. Some problems will include breakdowns when maintenance intervals exceed a certain level. This means that downtime may have to be factored in during the cycle. This will radically alter the results of the calculation. For example, continuing to expand the problem in Fig. 10-5, the downtime costs could be added with the result shown in Fig. 10-6. Now the downtime drives the true cost even higher than it was previously. However, in this example, the initial cost of the downtime is factored in immediately. The breakdown will

LOST PERFORMANCE CALCULATION

If the performance loss is linear, then at 4000 hours of operation, the loss is 5% and the value is $100.00, then the value of the loss is:

$$0.05 \times \$100.00/hr = \$5.00/hr$$

Therefore, at 4000 hours of operation, the pump is producing only a value of $95.00, or it is losing $5.00 per hour.

The table version would be:

Time Since Last Service (Hours)	Lost Performance Cost (Dollars)
100	0.13
500	0.63
1000	1.25
1500	1.88
2000	2.50
2500	3.12
3000	3.74
3500	4.36
4000	5.00

Figure 10-3. A simple method for calculating the performance falloff of a pump.

only occur if maintenance is not performed before 3000 hours of operation. If the maintenance frequency extends beyond that time, an additional cost of $2,400 (24 hours × $100 per hour value of the process) will be incurred. This removes any doubt that the lowest total cost would be around 2000 hours of operation, but definitely before 3000 hours of operation.

The ability to apply statistical techniques can go far beyond the simple example used here. Consider the ability to perform this type of cost analysis for each subcomponent of a large mechanical drive; the ability to determine the lowest life cycle cost for complex equipment; and the ability to determine the

TRUE LOST PERFORMANCE CALCULATION

The true lost performance cost is usually calculated by a calculus formula. However, a simple geometry formula can serve the purpose in the simple example.

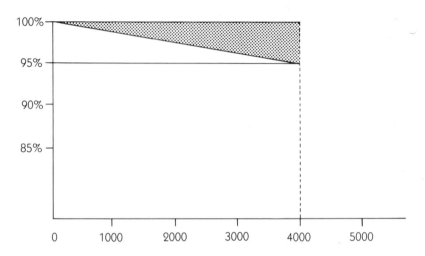

The total loss is not the entire rectangle, but only ½ of its area. So the loss would only be ½ of the calculated amount. The revised version of the table would be:

Time Since Last Service (Hours)	Lost Performance Cost (Dollars)
100	0.065
500	0.31
1000	0.63
1500	0.94
2000	1.25
2500	1.56
3000	1.87
3500	2.18
4000	2.50

Figure 10-4. Calculating the actual performance falloff of a pump.

TRUE TOTAL COST CALCULATION

The true total cost can now be established for this task. Combining the tables results in:

Time Since Last Service (Hours)	Maintenance Cost (Dollars)	Lost Performance Cost (Dollars)	True Total Costs (Dollars)
100	15.00	0.065	15.065
500	3.00	0.31	3.31
1000	1.50	0.63	2.13
1500	1.00	0.94	1.94
2000	0.75	1.25	2.00
2500	0.60	1.56	2.16
3000	0.50	1.87	2.37
3500	0.43	2.18	2.61
4000	0.38	2.50	2.88

Figure 10-5. A chart can be used to calculate the lowest true cost. This is the optimum opportunity for maintenance action.

amount of resources to be spent on redesign and retrofit engineering projects, based on anticipated return on the investment.

However, while this is a straightforward mathematical model, why are there so few companies doing this type of calculation? It is because very few of them have any reliable data with which to work. This is why the program outline for the American approach to T.P.M. places this step *after* good work order controls are implemented. The data to do these calculations must come from the work order system. Without these reliable data, companies will be back to guessing when maintenance activities should take place, and not optimizing their resources.

A second example of statistical controls for maintenance examines critical spares. Spare parts are an albatross for maintenance managers. The operations and upper management see critical spares (which are typically slow moving) as unnecessary or wasted, since they never seem to turn over. The same tech-

FACTORING IN BREAKDOWN COSTS

If a breakdown occurs and there is no spare, then the service also results in a cost penalty to the Operations Group.

— For planned service—8 hours of downtime
— If a breakdown occurs—24 hours of downtime

A breakdown will occur every 3000 hours of operation, based on repair records.

The table shows:

Time Since Last Service (Hours)	Maintenance Cost (Dollars)	Lost Performance Cost (Dollars)	Downtime Cost (Dollars)	True Total Costs (Dollars)
100	15.00	0.065	8.00	23.07
500	3.00	0.31	1.60	4.91
1000	1.50	0.63	0.80	2.93
1500	1.00	0.94	0.53	2.47
2000	0.75	1.25	0.40	2.40
2500	0.60	1.56	0.32	2.48
3000	0.50	1.87	1.07	3.44
3500	0.43	2.18	0.91	3.52
4000	0.38	2.50	0.80	3.68

Figure 10-6. Adding breakdown costs to the calculation provides a more realistic estimate of true total cost.

niques can be applied to the spares as were applied to the equipment. For example, consider what costs must go into a stocking decision, as follows.

How many are used per year, or "mean time between failures?"

What are the operating requirements for the equipment per year?

What is the cost of downtime if the part is unavailable?

What does the item cost?

What is the annual holding cost for the item?

How often can the item be repaired?

What is the lead time to get a replacement?

What is the lead time to repair the item?

After the costs related to the stocking decision are all identified, then a logical decision can be made concerning the stocking level. For example, the difference between keeping no spares and keeping one spare could be substantial when resulting downtime is factored in. Typically, critical spares have long lead times. If no spares are kept, the resulting downtime may be weeks of lost production. This can quickly add up to tens of thousands of dollars. This is a stiff penalty to pay for short-sighted decisions.

The same could be applied to normal stores items. If the same information is kept, statistical formulas can be used to calculate service levels, actual cost per item, number of projected turns per year, etc. These can provide a manager with the tools necessary to make accurate and cost-effective management decisions.

The drawback to all of the calculations is the need for accurate data. This is why the maintenance organization must have the discipline to collect accurate data before any of the techniques in this chapter can be used. Utilization of statistical techniques with poor or inaccurate data will provide just as inaccurate an answer as guessing at it. A good, disciplined approach to the work order system will enable many of the techniques mentioned here to be useful to the maintenance manager.

11 Team-Based Maintenance

Team-based maintenance is the step that all companies envision when they hear the term T.P.M. If the plan in this book has been followed to this point, what type of organizational practices are presently used? Up to this point, the organization should be experiencing World Class results. The organization is dedicated to using a disciplined approach to the work order system. All information is recorded completely and accurately through the work order system. The effective use of the work order system allows for accurate scheduling practices. Underlying the entire structure of the maintenance organization is the attitude of maintenance prevention. The preventive/predictive maintenance program is in an optimum condition. The costs related to P.M./ P.D.M. are calculated in the scope of the lowest total cost to the company.

These changes in the company show employees that the maintenance improvement program is no longer just another "program of the month." There has been a dedication to this program by management, and this dedication has been shown by the expenditure of resources to accomplish the improvements to date. With the time that has expired to bring the organization to this stage, the employee's attitude has become one of cooperation and understanding.

If, at this time, management begins the transfer of some basic maintenance tasks, such as routine inspections and service, the operations and maintenance groups will cooperate. This idea must be presented to the employees in the total cost concept. The benefits and results must be quantified to the employees. The objective of the program is not to eliminate any maintenance jobs; the true objective is to raise the level of maintenance on the equipment. If the operators are to inspect, clean, and routinely service their equipment, they will contribute to increased equipment effectiveness.

A good analogy to the new relationship between maintenance and operations is the relationship between paramedics and doctors. The paramedics are the first on the scene, applying first aid to the patient. If the patient revives, there is no need for the doctor. However, if there is a problem found with the patient, then the doctor (maintenance technician) needs to see the patient. The doctor can then perform any required maintenance tasks on the patient.

What is the biggest benefit to the company? In addition to seeing the equipment in a constantly improved condition, the typical minor stoppages are even further reduced. This will continue to increase the profit margin for the company, continuing to make it more competitive.

How is this program implemented? What are the program's limitations? What are the pitfalls that may be encountered? These are all questions that will be answered in the remainder of this chapter.

Operator-Based Maintenance

How is the program implemented? What are the steps that most companies can use to implement the program? They are as follows:

1. identifying the tasks to be transferred

2. training the operators for the tasks

3. monitoring the program.

The identification of the tasks to be transferred requires an understanding of what is being accomplished. The actual transfer of work must be in 10-minute blocks. If the operators are required to spend any more time performing maintenance tasks during a shift, their true jobs will suffer. The Japanese allow 10 minutes per shift and one 30-minute period some time during the week. If any more time than that is allotted to maintenance tasks, a proportionate decrease in production levels will be experienced.

The tasks that are to be transferred are: routine inspections, adjustments, and small services. Further description of the tasks includes the inspections for looseness, unusual wear, contamination, leaks, etc. The objective is not to have the operator fix all of the problems that are discovered, but to fix what he can and alert maintenance of the need for repair for the remainder of them. Whatever cannot be done by the operator in the allotted time must be performed on a planned and scheduled basis by the maintenance department. The adjustments will include tightening any loose parts to prevent vibration, ensuring the proper adjustment of all parts of the equipment the operator contacts, and the required lubrication of components.

How are these tasks identified? They are selected from the existing preventive maintenance program. If the present preventive maintenance tasks are properly detailed, then selecting the tasks to be transferred becomes simple. This process is one of the main reasons why the development of the preventive mainte- nance program in Chapter 5 is so important. By adjusting the present P.M. tasks into smaller blocks of time, they can easily be transferred to the operators without having to create a new P.M. program. Any tasks that are too complex or unsafe for the opera- tors to perform should remain part of the maintenance program.

Training the Operators

Some companies will think that the operators should already know how to perform the tasks relating to their equipment. This myth must be dispelled quickly. Just as all maintenance techni- cians will require training to do their tasks, so also will the operators. It will only take one missed problem on an inspection, one incorrect adjustment, or one missed lubrication point for a breakdown to occur. Over time, if breakdowns continue to occur, frustration will result for the operators and the mainte- nance technicians. This will lead to rapid deterioration of the entire program. It would be a crime to bring the organization this far and lose it because of failing to invest in operator training.

Another consideration for training the operators is safety. If the operators are not correctly trained, accidents will occur. This will allow the skeptics to point to problems within the program; and operators will only have so many accidents (as a group) before they will quit performing the tasks.

Since training is so important, where do you get the materials to train the operators? Chapter 9 detailed some maintenance training materials; it is merely a matter of adapting these mate- rials to the tasks that the operators now perform. This provides sufficient training to assure competency. Who does the training?

It is the maintenance technicians that work on the team with the operators. Operator training has to be like maintenance training—a mix of written instructions and actual monitored performance of the task. This also helps to provide a bonding between the operators and the maintenance technicians. Some companies even provide a certificate of completion each time an operator masters an assigned task.

The training must be complete since the operators must learn enough to set their own standards for

cleaning

lubrication

inspection

setups

adjustments

operational parameters

housekeeping.

Without adequate training, the continued reliance on the maintenance department will negate any advantages of going to operator-based maintenance. This is why the training of the technicians is equally important. If they are training the operators, they also must be technical enough to perform the above tasks. Sad to say, in most companies in 1991, very few technicians know enough to set maintenance standards in the above areas.

Monitoring the Program

Monitoring the program is a matter of measuring the effectiveness of the program. The equipment effectiveness is the ultimate measure of the program. If the equipment effectiveness is not improved, then the tasks being performed by the operators need examination. Are they adequate? Are they eliminating or preventing problems? If not, then appropriate steps need to be taken. The "continual and rapid improvement" which is the World Class theme must be employed in program monitoring. The answer to almost all problems in operator-based maintenance is training, and *more* training.

The transfer of some of the tasks to operations should in no way signal a defeat to maintenance. The next step for the maintenance organization is to begin the advanced training of the technicians. This allows the in-house technicians to take over the more advanced role of being a direct interface to engineering, outside vendors, and technical consultants. The transfer of the maintenance tasks to operations is limited. It is estimated that *approximately 20%* of the maintenance tasks (of the type mentioned previously) can be transferred to operations. This 20% should then be applied to learning to maintain and repair new technology. This allows the organization to continue to grow and become competitive.

The Manager's Role

Since the relationship between operations and maintenance changes under the T.P.M. banner, the manager's role must also change. When the operators and technicians work together, they will often form problem-solving teams. These teams may be for specific problems or may be used to monitor certain operational parameters. They work to solve the problems and generate improvements. In this manner, the control and decision making is pushed to the lowest possible level. This dramatically changes the manager's role from monitor to coach. In fact, the traditional management role evolves into a new World Class role. The role can be divided into the following four main areas:

recognize the importance of the work

help to set and achieve the goals

take action on the suggestions

reward the employee's efforts.

Recognizing the importance of the work is critical to the success of the program. If the employees do not feel that it is important, they will put little or no effort into the program. The supervisor's attitude will determine the importance the team places on the improvement effort. Remember: *if you don't think their efforts are important, they won't either.*

Helping the team set and achieve the goals means the supervisor will use knowledge of the corporate goals and objectives to steer the teams to work on problems that can support the corporate goals. Without this guidance, it is like running a race without a lane or finish line. With the supervisor providing some of the direction and helping to set goal values and a proposed time limit to the improvement project, the finish line is set. All humans work better when the goals and objectives are set. The trap here is to *steer* the group to this information, not to *dictate* it to them.

The supervisor next has the task of acting on any suggestions that the team develops. The supervisor must provide assistance in seeing that the suggestions are properly written, perhaps providing cost information or engineering data required to help substantiate the suggestion. The supervisor may even be called on to present the suggestion to upper management, if the team feels insecure in doing so. Once the plan has been presented, the supervisor can follow up to ensure implementation. This may be in the form of overcoming any problems or obstacles that develop.

Rewarding the team's efforts helps to satisfy an individual's desire for recognition. The rewards generally take the following three forms:

monetary rewards

plaques or certificates

formal notice presentations.

The supervisor must know the makeup of the team and which type of reward to use for which suggestion. The recognition will help to further motivate the team members to even more achievements. The recognition also helps to stimulate the creativity of other teams as well. The competitive situation helps to benefit the corporation, so when it is properly directed it can be very healthy.

As can be seen from the above descriptions, the supervisor's role changes from that of a monitor to an advisor or coach. This takes training and patience on the part of the supervisor. The supervisor's manager should be in a position to help make the changes required to manage in the new environment.

Steps to Autonomous Groups

What are autonomous groups? These are the teams of employees that work together, independent of direct supervisory control. The supervisors will function as described previously. However, the groups will continue to grow to maturity. They will progress through the following four distinct phases:

1. self-development

2. improvement activities

3. problem solving

4. autonomy.

The self-development phase is where the group begins to see the larger picture from the corporate position. They start seeing their role within the company and the resulting contributions that they can make. They must develop to the point where they learn new techniques and gather new information. They become proficient at understanding and applying the new information and techniques to their jobs. As they begin to mature as individuals, and as a group, they begin to understand the importance of each team member and the contributions that each member can make to the group. They learn to be comfortable working within the small group arrangement.

The improvement activities begin when the group starts to develop suggestions and ideas that are proposed and implemented. At this stage, the supervisors are still choosing the projects or areas that need improvement. When the suggestions or changes are implemented, they produce results. The results show improvement, which in turn produces a sense of accomplishment in the team members. This sense of accomplishment allows the team to take pride in their work, producing an even more intense desire to make additional contributions.

The problem-solving step is a selection process that allows the group to select from a list of problems. This allows them to find problems that are of particular importance to their departments or specific job duties. The independence of no longer being told which problem to work on is setting the stage for the development of autonomy. At this stage, the supervisor is in-

volved on an as-needed basis. The supervisor is no longer required to pick out the particular problem, since the group is
working off of a list. This phase is often difficult for the supervisors as their role begins to be altered dramatically.

The last phase is the autonomous group. This is where the
employees have matured to the point of requiring very little, if
any, direction. The teams have matured to the point of being able
to select goals that are consistent with corporate policies, and
they are fully capable of managing themselves. This form of
employee empowerment produces dramatic results within an
organization. With the decision making pushed down to the
lowest possible level, the theme of continuous and rapid improvement becomes a way of life, not just some words. The
organization that reaches this stage has truly become a World
Class company.

12
Measuring the Program Results

This chapter deals with measuring the results of a maintenance improvement program. However, a problem arises when trying to decide what maintenance indicators should be used. There are hundreds of so-called "indicators" that consultants will try to sell clients. While there are so many, the whole focus should be the financial bottom line. The total measure of maintenance performance is the percent of sales dollars spent on maintenance or, in the case of facilities, the maintenance cost per square foot maintained. Since one of the goals of World Class manufacturing or service is to provide all products and service at the lowest cost, this is where the rubber meets the road.

If a competitor has a maintenance cost equal to 4% of the sales dollar, and your maintenance cost is 7%, then the competitor enjoys a 3% advantage that he can use to convert to profit, reduce price, fund new capital investment, or put into research and development. In any case, it puts your company at a definite disadvantage.

What are some indicators that can be used, without entering into a "paralysis by analysis" type of operation? They can be divided into three main categories:

financial

effectiveness

performance (four levels).

The relationship between the various types of indicators is pictured in Fig. 12–1. Each level highlights an indicator. The cause of any problem, and its subsequent solution, is provided by examining the indicators in the next lower level. By the time any

MAINTENANCE PERFORMANCE MEASUREMENT PYRAMID

Figure 12-1. The maintenance performance measurement pyramid.

problem reaches the bottom level, either the supervisor, planner, or craft technicians should be able to provide a solution.

Financial indicators were discussed in Chapter 10. This is a different type of method than is traditionally used. However, in this area, we must get away from the "we have done it this way for years, why change" mindset. Only by examining the total maintenance cost can we become competitive. The top of the pyramid includes techniques that even the Japanese and other foreign competitors have not perfected.

The effectiveness of the maintenance organization (level 2) should always be measured by the equipment effectiveness formula. This formula highlights problems that could be misinterpreted to indicate maintenance problems. However, when the formulas that make up the equipment effectiveness formula are examined, the true source of the problem is uncovered.

What are some annual indicators that can be used to trend maintenance performance? They would include the following:

maintenance costs as a percentage of asset purchase price

maintenance costs as a percentage of estimated replacement value

maintenance costs as a percentage of total sales

maintenance costs as a percentage of square foot maintained.

While these indicators do not comprise all that might be selected, they are common indicators used for long-range trending.

The monthly indicators are those that need monitoring over a time period to trend effectiveness. They include the following.

Percentage of emergency work—hours charged to emergency work compared to the total maintenance hours expended.

Percentage of P.M./P.D.M. work—hours charged to P.M./P.D.M. compared to the total maintenance hours expended.

Percentage of downtime caused by breakdowns—hours of downtime caused by breakdowns compared to the total number of downtime hours.

Percentage of overhead—total dollars spent on maintenance overhead compared to the total maintenance labor dollars expended.

Percentage of maintenance labor and materials costs—the total labor cost compared to the total material cost.

Craft backlog—the amount of work ready to schedule in the craft backlog compared to the average amount of scheduled work completed each week.

Weekly maintenance performance indicators comprise the measurement of actual maintenance performance. This is the next step to the bottom level of the pyramid. Some of the common indicators are as follows.

Schedule compliance—number of work orders scheduled to be completed compared to the actual number of work orders completed.

Supervisor performance report—compares estimated labor and material requirements for scheduled jobs to the actual labor and material used.

Planner performance report—compares what was planned for the week versus what was actually completed.

Work order costs report—compares what was actually planned versus what costs were actually charged for the week's work by individual work orders.

Technician's time—what percentage of time was spent in emergency, preventive maintenance, normal scheduled, and outage work assignments.

Daily maintenance indicators are the day-to-day measures of maintenance activities. This is the bottom level of the pyramid. These are measures that should supply the information to feed the rest of the pyramid. They include the following.

Work completed—this compares the actual versus the estimate for each work order completed during the previous day.

Emergency work requests—a listing of all breakdown and critical work completed the day before.

Preventive maintenance work due and overdue—lists all P.M. work that is due and overdue as of the current date.

Stock replenishment list—a list of all stock items that have fallen below their minimum, and which needed reordering in the last 24 hours.

Technician's time—a list of each work order and the time spent on it by each technician for the last 24 hours.

It should be noted that this list of performance indicators is not complete. There are some organizations that will use additional or different indicators. This is just a list of suggested or common indicators. The important point is to select upwardly compatible performance indicators, so that the top of the pyramid has the necessary information to make the right decisions based on developed data.

What to Do with the Data

One of the most common mistakes is to neglect benchmarking the status of the maintenance organization before starting a

T.P.M. program. Since the steps outlined in this book include the vital step of benchmarking, you should have a frame of reference for this entire project. Benchmarking the results and then reporting on them is an important step in keeping management support for the program.

It is important to provide upper management with a report at least once every six months to update them on the status of the project. Any update should include the expenditures versus the benefits. It is realistic to expect to have to spend money at the start of the program before benefits are achieved. However, as long as this is according to the plan, there should be no problem on management's part. The problems occur when the results deviate from the plan. This is where the data you have been gathering can be used to explain the deviation, and also the correction, before the problem becomes critical.

If these techniques are used, it should simplify the effort required to convince management that maintenance is an important contributor to the corporate profit picture.

13 The Future of Maintenance

Where will maintenance go in the future? The answer lies in three basic areas:

management acceptance

technology

management techniques.

In the area of management acceptance, it is a matter of educating management as to the value of maintenance. Some of the techniques in the chapter on total cost will help. There is also another method. Study the history of the quality programs. There are amazing similarities between the obstacles that T.Q.C. programs had to overcome and the ones facing the maintenance organizations. Consider the works of Crosby, Deming, Juran, and others. Learn how they convinced management that it was time to take action. Then study the methodologies that they used to implement their ideas. These also will have major applications to the maintenance programs.

Consider Fig. 13–1. The pyramid in (A) is typical of American industries. They have tried to implement all of the various World Class programs, but have neglected to provide a good solid foundation (usually their assets). Over a period of time, the programs begin to crack and crumble. The shifting sands of the marketplace and competition have taken their toll. Then consider (B). A solid foundation of maintenance basics, coupled with World Class programs, presents a solid pyramid without any cracks. Since the assets will support the goals and objectives of the other programs, the company will remain competitive, no matter what is thrown against it.

It is only by convincing management of the true value of maintenance that any improvement can ever be made. Without convincing management, the funding and disciplines will never be provided.

In the area of technology, it is claimed that there is nothing new under the sun. In the last decade, there has been continual

WORLD CLASS PYRAMID

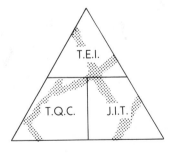 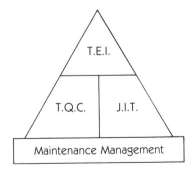

Figure 13-1. Maintenance management provides a sound foundation for industry.

refining of some technologies, but no new ones have been introduced. Predictive maintenance will continue to be emphasized, but new technologies will not be introduced to revolutionize this market.

If one area is going to be emphasized, it is the use of expert systems to help maintenance technicians be more effective in troubleshooting equipment. This technology, coupled with the real-time monitoring of data through PLC's, will enable the technicians to troubleshoot more accurately than ever before. The question will still revolve around the funding necessary to train the technicians to take advantage of these tools.

The final area to be considered is management techniques. After examining your maintenance supervisors and their skills, ask yourself: "Are they capable of managing in a World Class environment?" Most supervisors are 45 + years old, and many are hesitant and resistant to change. The techniques and skills required to manage in a T.P.M. environment will be entirely different. Will they be able to adapt?

In conclusion, maintenance is not hard to do, it is only hard to sell. The concepts and techniques that are being hailed as new and innovative are only variations of older techniques that never became popular when companies did not have to be competitive. The question now facing American managers is whether they want to change to be competitive, or not change and become extinct.

Index

A

B

C